Java并发实现原理
JDK源码剖析

余春龙 / 著

电子工业出版社
Publishing House of Electronics Industry
北京·BEIJING

内 容 简 介

本书全面而系统地剖析了 Java Concurrent 包中的每一个部分，对并发的实现原理进行了深入的探讨。全书分为 8 章，第 1 章从最基础的多线程知识讲起，厘清多线程中容易误解的知识点，探究背后的原理，包括内存重排序、happen-before、内存屏障等；第 2～8 章，从简单到复杂，逐个剖析 Concurrent 包的每个部分，包括原子类、锁、同步工具类、并发容器、线程池、ForkJoinPool 和 CompletableFuture。

本书适合有一定 Java 开发经验的工程师、架构师阅读。通过阅读本书，读者可以对多线程编程形成一个"深刻而直观"的认识，而不是仅仅停留在概念和理论层面。

未经许可，不得以任何方式复制或抄袭本书之部分或全部内容。
版权所有，侵权必究。

图书在版编目（CIP）数据

Java 并发实现原理：JDK 源码剖析 / 余春龙著.--北京：电子工业出版社，2020.4
ISBN 978-7-121-37972-7

Ⅰ．①J… Ⅱ．①余… Ⅲ．①JAVA 语言—程序设计Ⅳ．①TP312.8

中国版本图书馆 CIP 数据核字(2019)第 255587 号

责任编辑：宋亚东
印　　刷：北京盛通商印快线网络科技有限公司
装　　订：北京盛通商印快线网络科技有限公司
出版发行：电子工业出版社
　　　　　北京市海淀区万寿路 173 信箱　邮编：100036
开　　本：787×980　1/16　印张：16　字数：307.2 千字
版　　次：2020 年 4 月第 1 版
印　　次：2023 年 2 月第 6 次印刷
定　　价：89.00 元

凡所购买电子工业出版社图书有缺损问题，请向购买书店调换。若书店售缺，请与本社发行部联系，联系及邮购电话：(010) 88254888，88258888。

质量投诉请发邮件至 zlts@phei.com.cn，盗版侵权举报请发邮件至 dbqq@phei.com.cn。
本书咨询联系方式：(010) 51260888-819，faq@phei.com.cn。

前 言

并发编程可选择的方式有多进程、多线程和多协程。作者在另一本书《软件架构设计：大型网站技术架构与业务架构融合之道》中，曾对这三种方式进行了详细的比较。对于 Java 来说，它既不像 C++ 那样，在运行中调用 Linux 的系统 API 去 "fork" 出多个进程；也不像 Go 那样，在语言层面原生提供多协程。在 Java 中，并发就是多线程模式。

对于人脑的认知来说，"代码一行行串行"当然最容易理解。但在多线程下，多个线程的代码交叉并行，要访问互斥资源，要互相通信。作为开发者，需要仔细设计线程之间的互斥与同步，稍不留心，就会写出非线程安全的代码。正因此，多线程编程一直是一个被广泛而深入讨论的领域。

在 JDK 1.5 发布之前，Java 只在语言级别上提供一些简单的线程互斥与同步机制，也就是 synchronized 关键字、wait 与 notify。如果遇到复杂的多线程编程场景，就需要开发者基于这些简单的机制解决复杂的线程同步问题。而从 JDK 1.5 开始，并发编程大师 Doug Lea 奉上了一个系统而全面的并发编程框架——JDK Concurrent 包，里面包含了各种原子操作、线程安全的容器、线程池和异步编程等内容。

本书基于 JDK 7 和 JDK 8，对整个 Concurrent 包进行全面的源码剖析。JDK 8 中大部分并发功能的实现和 JDK 7 一样，但新增了一些额外特性。例如 CompletableFuture、ConcurrentHashMap 的新实现、StampedLock、LongAdder 等。

对整个 Concurrent 包的源码进行分析，有以下几个目的：

（1）帮助使用者合理地选择解决方案。Concurrent 包很庞大，有各式各样的线程互斥与同步机制。明白实现原理，使用者可以根据自己的业务场景，选择最适合自己的解决方案。避免重复造轮子，也避免因为使用不当而掉到"坑"里。

（2）对源码的分析，将让使用者对内存屏障、CAS 原子操作、锁、无锁等底层原理的认识，不再停留于一个"似是而非"的阶段，而是深刻地认识其本质。

（3）吸收借鉴大师的思维。在 Concurrent 包中，可以看到各种巧妙的并发处理策略。看了 Concurrent 包，才会发现在多线程中，不是只有简陋的互斥锁、通知机制和线程池。

本书将从多线程基础知识讲起，逐步地深入整个 Concurrent 包。读完本书，你将对多线程的原理、各种并发的设计原理有一个全面而深刻的理解。

限于时间和水平，书中难免有不足之处，望广大读者批评指正。

作　者

目 录

第 1 章 多线程基础 / 1

1.1 线程的优雅关闭 / 1
1.1.1 stop 与 destory 函数 / 1
1.1.2 守护线程 / 1
1.1.3 设置关闭的标志位 / 2

1.2 InterruptedException 与 interrupt()函数 / 3
1.2.1 什么情况下会抛出 Interrupted 异常 / 3
1.2.2 轻量级阻塞与重量级阻塞 / 4
1.2.3 t.isInterrupted()与 Thread.interrupted() 的区别 / 5

1.3 synchronized 关键字 / 5
1.3.1 锁的对象是什么 / 5
1.3.2 锁的本质是什么 / 6
1.3.3 synchronized 实现原理 / 7

1.4 wait 与 notify / 7
1.4.1 生产者-消费者模型 / 7
1.4.2 为什么必须和 synchronized 一起使用 / 8
1.4.3 为什么 wait()的时候必须释放锁 / 9
1.4.4 wait()与 notify()的问题 / 10

1.5 volatile 关键字 / 11
1.5.1 64 位写入的原子性（Half Write） / 11
1.5.2 内存可见性 / 11
1.5.3 重排序：DCL 问题 / 12

1.6 JMM 与 happen-before / 13
1.6.1 为什么会存在"内存可见性"问题 / 13
1.6.2 重排序与内存可见性的关系 / 15
1.6.3 as-if-serial 语义 / 16
1.6.4 happen-before 是什么 / 17
1.6.5 happen-before 的传递性 / 18
1.6.6 C++中的 volatile 关键字 / 19
1.6.7 JSR-133 对 volatile 语义的增强 / 20

1.7 内存屏障 / 20

1.7.1 Linux 中的内存屏障 / 21

1.7.2 JDK 中的内存屏障 / 23

1.7.3 volatile 实现原理 / 24

1.8 final 关键字 / 25

1.8.1 构造函数溢出问题 / 25

1.8.2 final 的 happen-before 语义 / 26

1.8.3 happen-before 规则总结 / 26

1.9 综合应用：无锁编程 / 27

1.9.1 一写一读的无锁队列：内存屏障 / 27

1.9.2 一写多读的无锁队列：volatile 关键字 / 27

1.9.3 多写多读的无锁队列：CAS / 28

1.9.4 无锁栈 / 28

1.9.5 无锁链表 / 28

第 2 章 Atomic 类 / 29

2.1 AtomicInteger 和 AtomicLong / 29

2.1.1 悲观锁与乐观锁 / 31

2.1.2 Unsafe 的 CAS 详解 / 31

2.1.3 自旋与阻塞 / 32

2.2 AtomicBoolean 和 AtomicReference / 33

2.2.1 为什么需要 AtomicBoolean / 33

2.2.2 如何支持 boolean 和 double 类型 / 33

2.3 AtomicStampedReference 和 AtomicMarkableReference / 34

2.3.1 ABA 问题与解决办法 / 34

2.3.2 为什么没有 AtomicStampedInteger 或 AtomictStampedLong / 35

2.3.3 AtomicMarkableReference / 36

2.4 AtomicIntegerFieldUpdater、AtomicLongFieldUpdater 和 AtomicReferenceFieldUpdater / 37

2.4.1 为什么需要 AtomicXXXFieldUpdater / 37

2.4.2 限制条件 / 38

2.5 AtomicIntegerArray、AtomicLongArray 和 AtomicReferenceArray / 38

2.5.1 使用方式 / 38

2.5.2 实现原理 / 39

2.6 Striped64 与 LongAdder / 40

2.6.1 LongAdder 原理 / 40

2.6.2 最终一致性 / 41

2.6.3 伪共享与缓存行填充 / 42

2.6.4 LongAdder 核心实现 / 43

2.6.5 LongAccumulator / 47

2.6.6 DoubleAdder 与 DoubleAccumulator / 47

第 3 章 Lock 与 Condition / 49

3.1 互斥锁 / 49

3.1.1 锁的可重入性 / 49

3.1.2 类继承层次 / 49

3.1.3 锁的公平性 vs.非公平性 / 51

3.1.4 锁实现的基本原理 / 51

3.1.5 公平与非公平的 lock()实现差异 / 53

3.1.6 阻塞队列与唤醒机制 / 55

3.1.7 unlock()实现分析 / 58

3.1.8 lockInterruptibly()实现分析 / 59

3.1.9 tryLock()实现分析 / 60

3.2 读写锁 / 60

3.2.1 类继承层次 / 60

3.2.2 读写锁实现的基本原理 / 61

3.2.3 AQS 的两对模板方法 / 62

3.2.4 WriteLock 公平 vs.非公平实现 / 65

3.2.5 ReadLock 公平 vs.非公平实现 / 67

3.3 Condition / 68

3.3.1 Condition 与 Lock 的关系 / 68

3.3.2 Condition 的使用场景 / 69

3.3.3 Condition 实现原理 / 71

3.3.4 await()实现分析 / 72

3.3.5 awaitUninterruptibly()实现分析 / 73

3.3.6 notify()实现分析 / 74

3.4 StampedLock / 75

3.4.1 为什么引入 StampedLock / 75

3.4.2 使用场景 / 75

3.4.3 "乐观读"的实现原理 / 77

3.4.4 悲观读/写:"阻塞"与"自旋"策略实现差异 / 78

第 4 章 同步工具类 / 83

4.1 Semaphore / 83

4.2 CountDownLatch / 84

4.2.1 CountDownLatch 使用场景 / 84

4.2.2 await()实现分析 / 85

4.2.3 countDown()实现分析 / 85

4.3 CyclicBarrier / 86

4.3.1 CyclicBarrier 使用场景 / 86

4.3.2 CyclicBarrier 实现原理 / 87

4.4 Exchanger / 90

4.4.1 Exchanger 使用场景 / 90

4.4.2 Exchanger 实现原理 / 91

4.4.3 exchange(V x)实现分析 / 92

4.5 Phaser / 94

4.5.1 用 Phaser 替代 CyclicBarrier 和 CountDownLatch / 94

4.5.2 Phaser 新特性 / 95

4.5.3 state 变量解析 / 96

4.5.4 阻塞与唤醒(Treiber Stack) / 98

4.5.5 arrive()函数分析 / 99

4.5.6 awaitAdvance()函数分析 / 101

第 5 章 并发容器 / 104

5.1 BlockingQueue / 104

5.1.1 ArrayBlockingQueue / 105

5.1.2 LinkedBlockingQueue / 106

5.1.3 PriorityBlockingQueue / 109

5.1.4 DelayQueue / 111

5.1.5 SynchronousQueue / 113

5.2 BlockingDeque / 121

5.3 CopyOnWrite / 123

 5.3.1 CopyOnWriteArrayList / 123

 5.3.2 CopyOnWriteArraySet / 124

5.4 ConcurrentLinkedQueue/Deque / 125

5.5 ConcurrentHashMap / 130

 5.5.1 JDK 7 中的实现方式 / 130

 5.5.2 JDK 8 中的实现方式 / 138

5.6 ConcurrentSkipListMap/Set / 152

 5.6.1 ConcurrentSkipListMap / 153

 5.6.2 ConcurrentSkipListSet / 162

第 6 章 线程池与 Future / 163

6.1 线程池的实现原理 / 163

6.2 线程池的类继承体系 / 164

6.3 ThreadPoolExecutor / 165

 6.3.1 核心数据结构 / 165

 6.3.2 核心配置参数解释 / 165

 6.3.3 线程池的优雅关闭 / 167

 6.3.4 任务的提交过程分析 / 172

 6.3.5 任务的执行过程分析 / 174

 6.3.6 线程池的 4 种拒绝策略 / 179

6.4 Callable 与 Future / 180

6.5 ScheduledThreadPoolExecutor / 183

 6.5.1 延迟执行和周期性执行的原理 / 184

 6.5.2 延迟执行 / 184

 6.5.3 周期性执行 / 185

6.6 Executors 工具类 / 188

第 7 章 ForkJoinPool / 190

7.1 ForkJoinPool 用法 / 190

7.2 核心数据结构 / 193

7.3 工作窃取队列 / 195

7.4 ForkJoinPool 状态控制 / 198

 7.4.1 状态变量 ctl 解析 / 198

 7.4.2 阻塞栈 Treiber Stack / 200

 7.4.3 ctl 变量的初始值 / 201

 7.4.4 ForkJoinWorkerThread 状态与个数分析 / 201

7.5 Worker 线程的阻塞-唤醒机制 / 202

 7.5.1 阻塞-入栈 / 202

 7.5.2 唤醒-出栈 / 204

7.6 任务的提交过程分析 / 205

 7.6.1 内部提交任务 pushTask / 206

 7.6.2 外部提交任务 addSubmission / 206

7.7 工作窃取算法：任务的执行过程分析 / 207

 7.7.1 顺序锁 SeqLock / 209

 7.7.2 scanGuard 解析 / 210

7.8 ForkJoinTask 的 fork/join / 212

7.8.1　fork　/ 213

7.8.2　join 的层层嵌套　/ 213

7.9　ForkJoinPool 的优雅关闭　/ 222

7.9.1　关键的 terminate 变量　/ 222

7.9.2　shutdown() 与 shutdownNow() 的区别　/ 223

第 8 章　CompletableFuture　/ 226

8.1　CompletableFuture 用法　/ 226

8.1.1　最简单的用法　/ 226

8.1.2　提交任务：runAsync 与 supplyAsync　/ 226

8.1.3　链式的 CompletableFuture：thenRun、thenAccept 和 thenApply　/ 227

8.1.4　CompletableFuture 的组合：thenCompose 与 thenCombine　/ 229

8.1.5　任意个 CompletableFuture 的组合　/ 231

8.2　四种任务原型　/ 233

8.3　CompletionStage 接口　/ 233

8.4　CompletableFuture 内部原理　/ 234

8.4.1　CompletableFuture 的构造：ForkJoinPool　/ 234

8.4.2　任务类型的适配　/ 235

8.4.3　任务的链式执行过程分析　/ 237

8.4.4　thenApply 与 thenApplyAsync 的区别　/ 241

8.5　任务的网状执行：有向无环图　/ 242

8.6　allOf 内部的计算图分析　/ 244

第 1 章 多线程基础

1.1 线程的优雅关闭

1.1.1 stop() 与 destory() 函数

线程是"一段运行中的代码",或者说是一个运行中的函数。既然是在运行中,就存在一个最基本的问题:运行到一半的线程能否强制杀死?

答案肯定是不能。在 Java 中,有 stop()、destory() 之类的函数,但这些函数都是官方明确不建议使用的。原因很简单,如果强制杀死线程,则线程中所使用的资源,例如文件描述符、网络连接等不能正常关闭。

因此,一个线程一旦运行起来,就不要去强行打断它,合理的关闭办法是让其运行完(也就是函数执行完毕),干净地释放掉所有资源,然后退出。如果是一个不断循环运行的线程,就需要用到线程间的通信机制,让主线程通知其退出。

1.1.2 守护线程

在下面的一段代码中:在 main(..) 函数中开了一个线程,不断循环打印。请问:main(..) 函数退出之后,该线程是否会被强制退出?整个进程是否会强制退出?

```
public static void main(String[] args) {
    System.out.println("main enter!");

    Thread t1 = new Thread(new Runnable() {
        @Override
        public void run() {
```

```
            while(true)
            {
                try{
                    System.out.println("t1 is executing");
                    Thread.sleep(500);
                }catch(InterruptedException e) {
                }
            }
        }
    });

    t1.start();
    System.out.println("main exit");   }
```

答案是不会的。在 C 语言中，main(..)函数退出后，整个程序也就退出了，但在 Java 中并非如此。

对于上面的程序，在 t1.start()前面加一行代码 t1.setDaemon(true)。当 main(..)函数退出后，线程 t1 就会退出，整个进程也会退出。

当在一个 JVM 进程里面开多个线程时，这些线程被分成两类：守护线程和非守护线程。默认开的都是非守护线程。在 Java 中有一个规定：当所有的非守护线程退出后，整个 JVM 进程就会退出。意思就是守护线程"不算作数"，守护线程不影响整个 JVM 进程的退出。例如，垃圾回收线程就是守护线程，它们在后台默默工作，当开发者的所有前台线程（非守护线程）都退出之后，整个 JVM 进程就退出了。

1.1.3 设置关闭的标志位

在上面的代码中，线程是一个死循环。但在实际工作中，开发人员通常不会这样编写，而是通过一个标志位来实现，如下面的代码所示。

代码 1

```
class MyThread extends Thread{
private boolean stopped = false;
    public void run()
{
    while(!stopped)
        {
            ...
        }
}
```

```
        public void stop(){
            this.stopped = true;
        }
    }
    public static void main(String[] args) {
        MyThread t = new MyThread();
            t.start();
            t.stop();  //通知线程 t 关闭
            t.join();  //等待线程 t 退出 while 循环，自行退出
    }
```

但上面的代码有一个问题：如果 MyThread t 在 while 循环中阻塞在某个地方，例如里面调用了 object.wait()函数，那它可能永远没有机会再执行 while(!stopped)代码，也就一直无法退出循环。

此时，就要用到下面所讲的 InterruptedException()与 interrupt()函数。

1.2 InterruptedException()函数与 interrupt()函数

1.2.1 什么情况下会抛出 Interrupted 异常

Interrupt 这个词很容易让人产生误解。从字面意思来看，好像是说一个线程运行到一半，把它中断了，然后抛出了 InterruptedException 异常，其实并不是。仍以上面的代码为例，假设 while 循环中没有调用任何的阻塞函数，就是通常的算术运算，或者打印一行日志，如下所示。

```
    public void run(){
        while(!stopped)
        {
            int a = 1, b=2;
            int c = a + b;
            System.out.println("thread is executing");
        }
    }
```

这个时候，在主线程中调用一句 t.interrupt()，请问该线程是否会抛出异常？答案是不会。

再举一个例子，假设这个线程阻塞在一个 synchronized 关键字的地方，正准备拿锁，如下代码所示。

```
    public void run(){
        while(!stopped)
        {
```

```
synchronized(obj1){          //线程阻塞在这个地方
    int a = 1, b=2;
    int c = a + b;
    System.out.println("thread is executing");
    }
  }
}
```

这个时候，在主线程中调用一句 t.interrupt()，请问该线程是否会抛出异常？答案是不会。

实际上，只有那些声明了会抛出 InterruptedException 的函数才会抛出异常，也就是下面这些常用的函数：

```
public static native void sleep(long millis) throws InterruptedException{…}
public final void wait() throws InterruptedException {…}
public final void join() throws InterruptedException{…}
```

1.2.2 轻量级阻塞与重量级阻塞

能够被中断的阻塞称为轻量级阻塞，对应的线程状态是 WAITING 或者 TIMED_WAITING；而像 synchronized 这种不能被中断的阻塞称为重量级阻塞，对应的状态是 BLOCKED。如图 1-1 所示的是在调用不同的函数之后，一个线程完整的状态迁移过程。

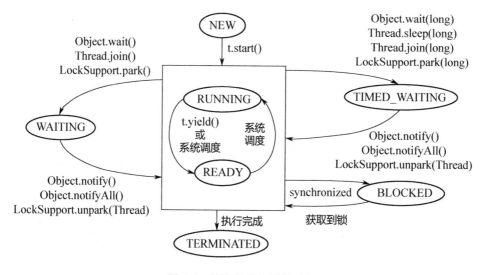

图 1-1 线程的状态迁移过程

初始线程处于 NEW 状态，调用 start() 之后开始执行，进入 RUNNING 或者 READY 状态。如果没有调用任何的阻塞函数，线程只会在 RUNNING 和 READY 之间切换，也就是系统的时间片调度。这两种状态的切换是操作系统完成的,开发者基本没有机会介入,除了可以调用 yield() 函数,放弃对 CPU 的占用。

一旦调用了图中的任何阻塞函数,线程就会进入 WAITING 或者 TIMED_WAITING 状态,两者的区别只是前者为无限期阻塞,后者则传入了一个时间参数,阻塞一个有限的时间。如果使用了 synchronized 关键字或者 synchronized 块,则会进入 BLOCKED 状态。

除了常用的阻塞/唤醒函数,还有一对不太常见的阻塞/唤醒函数,LockSupport.park()/unpark()。这对函数非常关键,Concurrent 包中 Lock 的实现即依赖这一对操作原语。

故而 t.interrupted() 的精确含义是"唤醒轻量级阻塞",而不是字面意思"中断一个线程"。

1.2.3　t.isInterrupted() 与 Thread.interrupted() 的区别

因为 t.interrupted() 相当于给线程发送了一个唤醒的信号,所以如果线程此时恰好处于 WAITING 或者 TIMED_WAITING 状态,就会抛出一个 InterruptedException,并且线程被唤醒。而如果线程此时并没有被阻塞,则线程什么都不会做。但在后续,线程可以判断自己是否收到过其他线程发来的中断信号,然后做一些对应的处理,这也是本节要讲的两个函数。

这两个函数都是线程用来判断自己是否收到过中断信号的,前者是非静态函数,后者是静态函数。二者的区别在于,前者只是读取中断状态,不修改状态;后者不仅读取中断状态,还会重置中断标志位。

1.3　synchronized 关键字

1.3.1　锁的对象是什么

对不熟悉多线程原理的人来说,很容易误解 synchronized 关键字:它通常加在所有的静态成员函数和非静态成员函数的前面,表面看好像是"函数之间的互斥",其实不是。synchronized 关键字其实是"给某个对象加了把锁",这个锁究竟加在了什么对象上面?如下面的代码所示,给函数 f1()、f2() 加上 synchronized 关键字。

```
class A {
   public void synchronized f1() {…}
   public static void synchronized f2() {…}
}
```

等价于如下代码:

```
class A {
    public void f1() {
        synchronized(this) {…}
    }
    public static void f2() {
        synchronized(A.class) {…}
    }
}
A a = new A()
a.f1();
a.f2();
```

对于非静态成员函数，锁其实是加在对象 a 上面的；对于静态成员函数，锁是加在 A.class 上面的。当然，class 本身也是对象。

这间接回答了关于 synchronized 的常见问题：一个静态成员函数和一个非静态成员函数，都加了 synchronized 关键字，分别被两个线程调用，它们是否互斥？很显然，因为是两把不同的锁，所以不会互斥。

1.3.2 锁的本质是什么

无论使用什么编程语言，只要是多线程的，就一定会涉及锁。既然锁如此常见，那么锁的本质到底是什么呢？

如图 1-2 所示，多个线程要访问同一个资源。线程就是一段段运行的代码；资源就是一个变量、一个对象或一个文件等；而锁就是要实现线程对资源的访问控制，保证同一时间只能有一个线程去访问某一个资源。打个比方，线程就是一个个游客，资源就是一个待参观的房子。这个房子同一时间只允许一个游客进去参观，当一个人出来后下一个人才能进去。而锁，就是这个房子门口的守卫。如果同一时间允许多个游客参观，锁就变成信号量，这点后面会专门讨论。

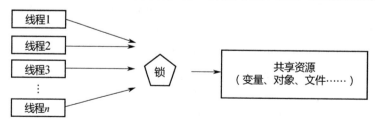

图 1-2 线程、锁和资源三者关系示意图

从程序角度来看，锁其实就是一个"对象"，这个对象要完成以下几件事情：

（1）这个对象内部得有一个标志位（state 变量），记录自己有没有被某个线程占用（也就是记录当前有没有游客已经进入了房子）。最简单的情况是这个 state 有 0、1 两个取值，0 表示没有线程占用这个锁，1 表示有某个线程占用了这个锁。

（2）如果这个对象被某个线程占用，它得记录这个线程的 thread ID，知道自己是被哪个线程占用了（也就是记录现在是谁在房子里面）。

（3）这个对象还得维护一个 thread id list，记录其他所有阻塞的、等待拿这个锁的线程（也就是记录所有在外边等待的游客）。在当前线程释放锁之后（也就是把 state 从 1 改回 0），从这个 thread id list 里面取一个线程唤醒。

既然锁是一个"对象"，要访问的共享资源本身也是一个对象，例如前面的对象 a，这两个对象可以合成一个对象。代码就变成 synchronized(this) {…}，我们要访问的共享资源是对象 a，锁也是加在对象 a 上面的。当然，也可以另外新建一个对象，代码变成 synchronized(obj1) {…}。这个时候，访问的共享资源是对象 a，而锁是加在新建的对象 obj1 上面的。

资源和锁合二为一，使得在 Java 里面，synchronized 关键字可以加在任何对象的成员上面。这意味着，这个对象既是共享资源，同时也具备"锁"的功能！

下面来看 Java 是如何做到让任何一个对象都具备"锁"的功能的，这也就是 synchronized 的实现原理。

1.3.3 synchronized 实现原理

答案在 Java 的对象头里。在对象头里，有一块数据叫 Mark Word。在 64 位机器上，Mark Word 是 8 字节（64 位）的，这 64 位中有 2 个重要字段：锁标志位和占用该锁的 thread ID。因为不同版本的 JVM 实现，对象头的数据结构会有各种差异，此处不再进一步讨论。

此处主要是想说明锁实现的思路，因为后面讲 ReentrantLock 的详细实现时，也基于类似的思路。在这个基本的思路之上，synchronized 还会有偏向、自旋等优化策略，ReentrantLock 同样会用到这些优化策略，到时会结合代码详细展开。

1.4 wait() 与 notify()

1.4.1 生产者-消费者模型

生产者-消费者模型是一个常见的多线程编程模型，如图 1-3 所示。

一个内存队列，多个生产者线程往内存队列中放数据；多个消费者线程从内存队列中取数据。要实现这样一个编程模型，需要做下面几件事情：

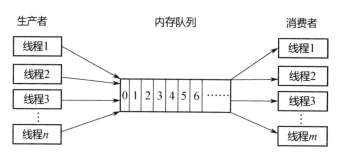

图 1-3　生产者-消费者模型

（1）内存队列本身要加锁，才能实现线程安全。

（2）阻塞。当内存队列满了，生产者放不进去时，会被阻塞；当内存队列是空的时候，消费者无事可做，会被阻塞。

（3）双向通知。消费者被阻塞之后，生产者放入新数据，要 notify() 消费者；反之，生产者被阻塞之后，消费者消费了数据，要 notify() 生产者。

第（1）件事情必须要做，第（2）件和第（3）件事情不一定要做。例如，可以采取一个简单的办法，生产者放不进去之后，睡眠几百毫秒再重试，消费者取不到数据之后，睡眠几百毫秒再重试。但这个办法效率低下，也不实时。所以，我们只讨论如何阻塞、如何通知的问题。

1. 如何阻塞？

办法 1：线程自己阻塞自己，也就是生产者、消费者线程各自调用 wait() 和 notify()。

办法 2：用一个阻塞队列，当取不到或者放不进去数据的时候，入队/出队函数本身就是阻塞的。这也就是 BlockingQueue 的实现，后面会详细讲述。

2. 如何双向通知？

办法 1：wait() 与 notify() 机制。

办法 2：Condition 机制。

此处，先讲 wait() 与 notify() 机制，后面会专门讲 Condition 机制与 BlockingQueue 机制。

1.4.2　为什么必须和 synchronized 一起使用

在 Java 里面，wait() 和 notify() 是 Object 的成员函数，是基础中的基础。为什么 Java 要把 wait() 和 notify() 放在如此基础的类里面，而不是作为像 Thread 一类的成员函数，或者其他类的成员函数呢？

在回答这个问题之前，先要回答为什么 wait() 和 notify() 必须和 synchronized 一起使用？请看下面的代码：

```
class A {
  private Object obj1 = new Object();
  public void f1() {
    synchronized(obj1) {
      ...
      obj1.wait()
      ...
    }
  }
  public void f2() {
    synchronized(obj1) {
      ...
      obj1.notify()
      ...
    }
  }
}
```

或者下面的代码：

```
class A {
  public void synchronized f1() {
    ...
    this.wait()
    ...
  }
  public void synchronized f2() {
    ...
    this.notify()
    ...
  }
}
```

然后，开两个线程，线程 A 调用 f1()，线程 B 调用 f2()。答案已经很明显：两个线程之间要通信，对于同一个对象来说，一个线程调用该对象的 wait()，另一个线程调用该对象的 notify()，该对象本身就需要同步！所以，在调用 wait()、notify() 之前，要先通过 synchronized 关键字同步给对象，也就是给该对象加锁。

前面已经讲了，synchronized 关键字可以加在任何对象的成员函数上面，任何对象都可能成为锁。那么，wait() 和 notify() 要同样如此普及，也只能放在 Object 里面了。

1.4.3　为什么 wait() 的时候必须释放锁

当线程 A 进入 synchronized(obj1) 中之后，也就是对 obj1 上了锁。此时，调用 wait() 进入阻塞状态，一直不能退出 synchronized 代码块；那么，线程 B 永远无法进入 synchronized(obj1)同步块里，永远没有机会调用 notify()，岂不是死锁了？

这就涉及一个关键的问题：在 wait()的内部，会先释放锁 obj1，然后进入阻塞状态，之后，它被另外一个线程用 notify()唤醒，去重新拿锁！其次，wait()调用完成后，执行后面的业务逻辑代码，然后退出 synchronized 同步块，再次释放锁。

wait()内部的伪代码如下：

```
wait() {
//释放锁
//阻塞，等待被其他线程notify
//重新拿锁
}
```

只有如此，才能避免上面所说的死锁问题。后面讲 Condition 实现的时候，会再详细讨论这个问题。

1.4.4 wait() 与 notify() 的问题

以上述的生产者-消费者模型来看，其伪代码大致如下：

```
public void enqueue() {
  synchronized(queue) {
    while(queue.full())  queue.wait();
    …. //入队列
    queue.notify()   //通知消费者，队列中有数据了
  }
}
public void dequeue() {
  synchronized(queue) {
    while(queue.empty())  queue.wait();
    …. //出队列
    queue.notify()   //通知生产者，队列中有空位了，可以继续放数据
  }
}
```

生产者本来只想通知消费者，但它把其他的生产者也通知了；消费者本来只想通知生产者，但它被其他的消费者通知了。原因就是 wait()和 notify()所作用的对象和 synchronized 所作用的对象是同一个，只能有一个对象，无法区分队列空和列队满两个条件。这正是 Condition 要解决的问题。

1.5 volatile 关键字

volatile 这个关键字很不起眼，其使用场景和语义不像 synchronized、wait() 和 notify() 那么明显。正因为其隐晦，volatile 关键字可能是在多线程编程领域中被误解最多的一个。而关键字越隐晦，背后隐含的含义往往越复杂、越深刻。接下来的几个小节将一步步由浅入深地从使用场景讨论到其底层的实现。

1.5.1 64 位写入的原子性（Half Write）

举一个简单的例子，对于一个 long 型变量的赋值和取值操作而言，在多线程场景下，线程 A 调用 set(100)，线程 B 调用 get()，在某些场景下，返回值可能不是 100。

```
public class Example1
{
  private long a = 0;
  public void set(long a)           //线程A调用set(100)
  {
     this.a = a;
  }
  public long get()                 //线程B调用get()，返回值是不是一定为100？
  {
     return this.a;
  }
}
```

这有点反直觉，如此简单的一个赋值和取值操作，在多线程下面为什么会不对呢？这是因为 JVM 的规范并没有要求 64 位的 long 或者 double 的写入是原子的。在 32 位的机器上，一个 64 位变量的写入可能被拆分成两个 32 位的写操作来执行。这样一来，读取的线程就可能读到 "一半的值"。解决办法也很简单，在 long 前面加上 volatile 关键字。

1.5.2 内存可见性

不仅 64 位，32 位或者位数更小的赋值和取值操作，其实也有问题。以 1.1 节中，线程关闭的标志位 stopped 为例，它是一个 boolean 类型的数字，也可能出现主线程把它设置成 true，而工作线程读到的却还是 false 的情形，这就更反直觉了。

注意，这里并不是说永远读到的都是 false，而是说一个线程写完之后，另外一个线程立即去读，读到的是 false，但之后能读到 true，也就是 "最终一致性"，不是 "强一致性"。这种特性，对于 1.1 节中的例子而言并没有太大影响，但如果想实现无锁算法，例如实现一把自旋锁，就会出现一个线程把状态置为了 true，另外一个线程读到的却还是 false，然后两个线程都会拿到这把锁的问题。

所以，我们所说的"内存可见性"，指的是"写完之后立即对其他线程可见"，它的反面不是"不可见"，而是"稍后才可见"。解决这个问题很容易，给变量加上 volatile 关键字即可。

"内存可见性"问题的出现，跟现代 CPU 的架构密切相关，1.6 节会详细探讨。

1.5.3　重排序：DCL 问题

单例模式的线程安全的写法不止一种，常用写法为 DCL（Double Checking Locking），如下所示。

```
public class Sington
{
private static Sington instance;
public static Sington getInstance()
{
  if(instance == null)                    //DCL
  {
    synchronized(Sington.class)           //为了性能，延迟使用 synchronized
    {
      if(instance == null)
         instance = new Instance();       //有问题的代码！
    }
  }
  return instance;
}
```

上述的 instance = new Instance() 代码有问题：其底层会分为三个操作：

（1）分配一块内存。

（2）在内存上初始化成员变量。

（3）把 instance 引用指向内存。

在这三个操作中，操作（2）和操作（3）可能重排序，即先把 instance 指向内存，再初始化成员变量，因为二者并没有先后的依赖关系。此时，另外一个线程可能拿到一个未完全初始化的对象。这时，直接访问里面的成员变量，就可能出错。这就是典型的"构造函数溢出"问题。解决办法也很简单，就是为 instance 变量加上 volatile 修饰。

通过上面的例子，可以总结出 volatile 的三重功效：64 位写入的原子性、内存可见性和禁止重排序。接下来，我们进入 volatile 原理的探究。

1.6　JMM 与 happen-before

1.6.1　为什么会存在"内存可见性"问题

要解释清楚这个问题，就涉及现代 CPU 的架构。图 1-4 所示为 x86 架构下 CPU 缓存的布局，即在一个 CPU 4 核下，L1、L2、L3 三级缓存与主内存的布局。每个核上面有 L1、L2 缓存，L3 缓存为所有核共用。

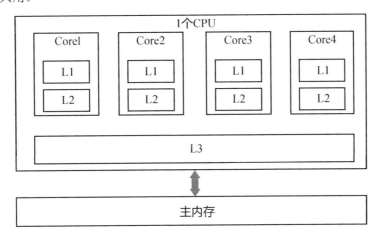

图 1-4　x86 架构下 CPU 缓存布局

因为存在 CPU 缓存一致性协议，例如 MESI，多个 CPU 之间的缓存不会出现不同步的问题，不会有"内存可见性"问题。

但是，缓存一致性协议对性能有很大损耗，为了解决这个问题，CPU 的设计者们在这个基础上又进行了各种优化。例如，在计算单元和 L1 之间加了 Store Buffer、Load Buffer（还有其他各种 Buffer），如图 1-5 所示。

L1、L2、L3 和主内存之间是同步的，有缓存一致性协议的保证，但是 Store Buffer、Load Buffer 和 L1 之间却是异步的。也就是说，往内存中写入一个变量，这个变量会保存在 Store Buffer 里面，稍后才异步地写入 L1 中，同时同步写入主内存中。

注意，这里只是简要画了 x86 的 CPU 缓存体系，还没有进一步讨论 SMP 架构和 NUMA 的区别，还有其他 CPU 架构体系，例如 PowerPC、MIPS、ARM 等，不同 CPU 的缓存体系会有各种差异。

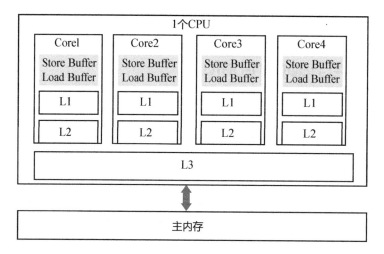

图 1-5　加了 Store Buffer 和 Load Buffer 的 CPU 缓存体系

但站在操作系统内核的角度，可以统一看待这件事情，也就是图 1-6 所示的操作系统内核视角下的 CPU 缓存模型。

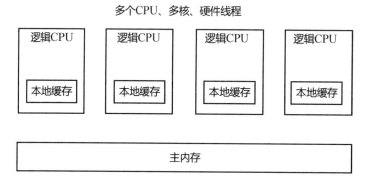

图 1-6　操作系统内核视角下的 CPU 缓存模型

多 CPU，每个 CPU 多核，每个核上面可能还有多个硬件线程，对于操作系统来讲，就相当于一个个的逻辑 CPU。每个逻辑 CPU 都有自己的缓存，这些缓存和主内存之间不是完全同步的。

对应到 Java 里，就是 JVM 抽象内存模型，如图 1-7 所示。

到此为止，介绍了不同 CPU 架构下复杂的缓存体系，也就回答了为什么会出现"内存可见性"问题。

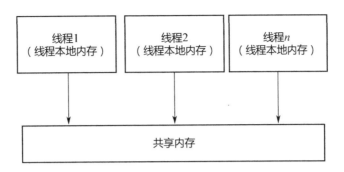

图 1-7　JVM 抽象内存模型

1.6.2　重排序与内存可见性的关系

Store Buffer 的延迟写入是重排序的一种，称为内存重排序（Memory Ordering）。除此之外，还有编译器和 CPU 的指令重排序。下面对重排序做一个分类：

（1）编译器重排序。对于没有先后依赖关系的语句，编译器可以重新调整语句的执行顺序。

（2）CPU 指令重排序。在指令级别，让没有依赖关系的多条指令并行。

（3）CPU 内存重排序。CPU 有自己的缓存，指令的执行顺序和写入主内存的顺序不完全一致。

在三种重排序中，第三类就是造成"内存可见性"问题的主因，下面再举一个例子来进一步说明这个问题。如下所示。

线程 1：

$X=1$

$a=Y$

线程 2：

$Y=1$

$b=X$

假设 X、Y 是两个全局变量，初始的时候，$X=0$，$Y=0$。请问，这两个线程执行完毕之后，a、b 的正确结果应该是什么？

很显然，线程 1 和线程 2 的执行先后顺序是不确定的，可能顺序执行，也可能交叉执行，最终正确的结果可能是：

（1）$a = 0$，$b = 1$

（2）$a=1$，$b=0$

（3）$a=1$，$b=1$

也就是不管谁先谁后，执行结果应该是这三种场景中的一种。但实际可能是 $a=0$，$b=0$。

两个线程的指令都没有重排序，执行顺序就是代码的顺序，但仍然可能出现 $a=0$，$b=0$。原因是线程 1 先执行 $X=1$，后执行 $a=Y$，但此时 $X=1$ 还在自己的 Store Buffer 里面，没有及时写入主内存中。所以，线程 2 看到的 X 还是 0。线程 2 的道理与此相同。

这就是一个有意思的地方，虽然线程 1 觉得自己是按代码顺序正常执行的，但在线程 2 看来，$a=Y$ 和 $X=1$ 顺序却是颠倒的。指令没有重排序，是写入内存的操作被延迟了，也就是内存被重排序了，这就造成内存可见性问题。

1.6.3 as-if-serial 语义

对开发者而言，当然希望不要有任何的重排序，这样理解起来最简单，指令执行顺序和代码顺序严格一致，写内存的顺序也严格地和代码顺序一致。但是，从编译器和 CPU 的角度来看，希望尽最大可能进行重排序，提升运行效率。于是，问题就来了，重排序的原则是什么？什么场景下可以重排序，什么场景下不能重排序呢？

1. 单线程程序的重排序规则

无论什么语言，站在编译器和 CPU 的角度来说，不管怎么重排序，单线程程序的执行结果不能改变，这就是单线程程序的重排序规则。换句话说，只要操作之间没有数据依赖性，如上例所示，编译器和 CPU 都可以任意重排序，因为执行结果不会改变，代码看起来就像是完全串行地一行行从头执行到尾，这也就是 as-if-serial 语义。对于单线程程序来说，编译器和 CPU 可能做了重排序，但开发者感知不到，也不存在内存可见性问题。

2. 多线程程序的重排序规则

编译器和 CPU 的这一行为对于单线程程序没有影响，但对多线程程序却有影响。对于多线程程序来说，线程之间的数据依赖性太复杂，编译器和 CPU 没有办法完全理解这种依赖性并据此做出最合理的优化。所以，编译器和 CPU 只能保证每个线程的 as-if-serial 语义。线程之间的数据依赖和相互影响，需要编译器和 CPU 的上层来确定。上层要告知编译器和 CPU 在多线程场景下什么时候可以重排序，什么时候不能重排序。

如图 1-8 所示，编译器和 CPU 遵守了 as-if-serial 语义，保证每个线程内部都是"看似完全串行的"。但多个线程会互相读取和写入共享的变量，对于这种相互影响，编译器和 CPU 不会考虑。

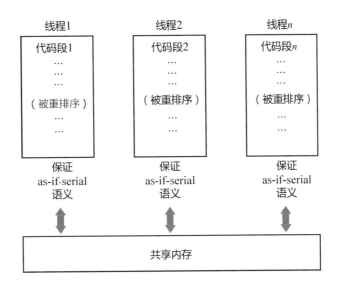

图 1-8 编译器和 CPU 保证每个线程的 as-if-serial 语义

1.6.4 happen-before 是什么

为了明确定义在多线程场景下，什么时候可以重排序，什么时候不能重排序，Java 引入了 JMM（Java Memory Model），也就是 Java 内存模型（单线程场景不用说明，有 as-if-serial 语义保证）。这个模型就是一套规范，对上，是 JVM 和开发者之间的协定；对下，是 JVM 和编译器、CPU 之间的协定。

定义这套规范，其实是要在开发者写程序的方便性和系统运行的效率之间找到一个平衡点。一方面，要让编译器和 CPU 可以灵活的重排序；另一方面，要对开发者做一些承诺，明确告知开发者不需要感知什么样的重排序，需要感知什么样的重排序。然后，根据需要决定这种重排序对程序是否有影响。如果有影响，就需要开发者显示地通过 volatile、synchronized 等线程同步机制来禁止重排序。

为了描述这个规范，JMM 引入了 happen-before，使用 happen-before 描述两个操作之间的内存可见性。那么，happen-before 是什么呢？

如果 A happen-before B，意味着 A 的执行结果必须对 B 可见，也就是保证跨线程的内存可见性。A happen before B 不代表 A 一定在 B 之前执行。因为，对于多线程程序而言，两个操作的执行顺序是不确定的。happen-before 只确保如果 A 在 B 之前执行，则 A 的执行结果必须对 B 可见。定义了内存可见性的约束，也就定义了一系列重排序的约束。

基于 happen-before 的这种描述方法，JMM 对开发者做出了一系列承诺：

（1）单线程中的每个操作，happen-before 对应该线程中任意后续操作（也就是 as-if-serial 语义保证）。

（2）对 volatile 变量的写入，happen-before 对应后续对这个变量的读取。

（3）对 synchronized 的解锁，happen-before 对应后续对这个锁的加锁。

……

对于非 volatile 变量的写入和读取，不在这个承诺之列。通俗来讲，就是 JMM 对编译器和 CPU 来说，volatile 变量不能重排序；非 volatile 变量可以任意重排序。JMM 没有对非 volatile 变量做这个承诺，所以出现了前面例子中的各种问题。

1.6.5 happen-before 的传递性

除了这些基本的 happen-before 规则，happen-before 还具有传递性，即若 A happen-before B，B happen-before C，则 A happen-before C。

如果一个变量不是 volatile 变量，当一个线程读取、一个线程写入时可能有问题。那岂不是说，在多线程程序中，我们要么加锁，要么必须把所有变量都声明为 volatile 变量？这显然不可能，而这就得归功于 happen-before 的传递性。

来看下面的例子：

```
class A {
private int a = 0;
private volatile int c = 0;
public void set() {
    a = 5;  //操作1
    c = 1;  //操作2
  }
 public int get(){
  int d = c;    //操作3
  return a;     //操作4
  }
}
```

假设线程 A 先调用了 set，设置了 a = 5；之后线程 B 调用了 get，返回值一定是 a = 5。为什么呢？

操作 1 和操作 2 是在同一个线程内存中执行的，操作 1 happen-before 操作 2，同理，操作 3 happen-before 操作 4。又因为 c 是 volatile 变量，对 c 的写入 happen-before 对 c 的读取，所以操作 2 happen-before 操作 3。利用 happen-before 的传递性，就得到：

操作 1 happen-before 操作 2 happen-before 操作 3 happen-before 操作 4。

所以，操作 1 的结果，一定对操作 4 可见。

再看一个例子：
```
class A {
  private int a = 0;
  private int c = 0;
  public synchronized void set() {
     a = 5;       //操作1
     c = 1;       //操作2
  }
  public synchronized int get(){
     return a;
  }
}
```
假设线程 A 先调用了 set，设置了 a = 5；之后线程 B 调用了 get，返回值也一定是 a = 5。

因为与 volatile 一样，synchronized 同样具有 happen-before 语义。展开上面的代码可得到类似于下面的伪代码：

```
线程A：
   加锁；        //操作1
   a = 5;       //操作2
   c = 1;       //操作3
   解锁；        //操作4
线程B：
   加锁；        //操作5
   读取 a；      //操作6
   解锁；        //操作7
```

根据 synchronized 的 happen-before 语义，操作 4 happen-before 操作 5，再结合传递性，最终就会得到：

操作 1 happen-before 操作 2 …… happen-before 操作 7。所以，a、c 都不是 volatile 变量，但仍然有内存可见性。

happen-before 的传递性非常有用，后面讲到 Concurrent 包的很多实现的时候，还会用到这个特性。

1.6.6　C++中的 volatile 关键字

在 C++中也有 volatile 关键字，但其含义和 Java 中的有一些差别。考虑下面的代码：

```
class A {
private volatile boolean done = false;
 public void process() {
     doSomeThing()        //做某些复杂的业务逻辑
     done = true;         //错误地方。done = true 和 doSomeThing()可能被重排序
  }
  public void afterProcess(){
    if(done) {
}
}
}
```

通过 done 这个标志位来标志一个任务是否完成。假设一个线程 A 调用 process，另一个线程不断地轮询调用 afterProcess，当 done = true 的时候，做一些额外工作。

这个代码在 C++中是错误的，但在 Java 中却是正确的。因为 Java 中的 volatile 关键字不仅具有内存可见性，还会禁止 volatile 变量写入和非 volatile 变量写入的重排序，但 C++中的 volatile 关键字不会禁止这种重排序。

1.6.7　JSR-133 对 volatile 语义的增强

Java 的 volatile 比 C++多出的这点特性，正是 JSR-133 对 volatile 语义的增强。下面这段话摘自 JSR-133 的原文：

What was wrong with the old memory model?
The old memory model allowed for volatile writes to be reordered with nonvolatile reads and writes, which was not consistent with most developers intuitions about volatile and therefore caused confusion.

也就是说，在旧的 JMM 模型中，volatile 变量的写入会和非 volatile 变量的读取或写入重排序，正如 C++中所做的。但新的模型不会，这也正体现了 Java 对 happen-before 规则的严格遵守。

1.7　内存屏障

为了禁止编译器重排序和 CPU 重排序，在编译器和 CPU 层面都有对应的指令，也就是内存屏障（Memory Barrier）。这也正是 JMM 和 happen-before 规则的底层实现原理。

编译器的内存屏障，只是为了告诉编译器不要对指令进行重排序。当编译完成之后，这种内存屏障就消失了，CPU 并不会感知到编译器中内存屏障的存在。

而 CPU 的内存屏障是 CPU 提供的指令，可以由开发者显示调用。下面主要讲 CPU 的内存屏障。

1.7.1　Linux 中的内存屏障

为了直观地说明内存屏障，下面摘录了 Linux 内核 kfifo.c 的源代码。

这是一个 RingBuffer，允许一个线程写入，一个线程读取（只能一写一读），整个代码没加任何的锁，也没有 CAS，但是线程是安全的，这是如何做到的呢？

```c
//入队（将数据插入 Ringbuffer）
unsigned int __kfifo_in(struct __kfifo *fifo,
        const void *buf, unsigned int len)
{
    unsigned int l;
    l = kfifo_unused(fifo);
    if (len > l)
        len = l;
    kfifo_copy_in(fifo, buf, len, fifo->in);
    //在这插入的是 Store Barrier 屏障
    fifo->in += len;
    return len;
}
static void kfifo_copy_in(struct __kfifo *fifo, const void *src,
        unsigned int len, unsigned int off)
{
    unsigned int size = fifo->mask + 1;
    unsigned int esize = fifo->esize;
    unsigned int l;
    off &= fifo->mask;
    if (esize != 1) {
        off *= esize;
        size *= esize;
        len *= esize;
    }
    l = min(len, size - off);
    memcpy(fifo->data + off, src, l);
    memcpy(fifo->data, src + l, len - l);

    //关键点：插入了一个 Store Barrier。从而保证先插入数据，再更新指针 in
    smp_wmb();
}

//出队（从 Ringbuffer 取出数据）
unsigned int __kfifo_out(struct __kfifo *fifo,
```

```c
                                                    void    *buf,
unsigned int len)
{
    len = __kfifo_out_peek(fifo, buf, len);
    //在这插入的是 Store Barrier 屏障
    fifo->out += len;
    return len;
}
unsigned int __kfifo_out_peek(struct __kfifo *fifo,
        void *buf, unsigned int len)
{
    unsigned int l;
    l = fifo->in - fifo->out;
    if (len > l)
        len = l;
    kfifo_copy_out(fifo, buf, len, fifo->out);
    return len;
}
static void kfifo_copy_out(struct __kfifo *fifo, void *dst,
        unsigned int len, unsigned int off)
{
    unsigned int size = fifo->mask + 1;
    unsigned int esize = fifo->esize;
    unsigned int l;
    off &= fifo->mask;
    if (esize != 1) {
        off *= esize;
        size *= esize;
        len *= esize;
    }
    l = min(len, size - off);
    memcpy(dst, fifo->data + off, l);
    memcpy(dst + l, fifo->data, len - l);
    //关键点：插入了一个 Store Barrier。从而保证先出队，再更新指针 out
    smp_wmb();
}
```

通过上面代码可以看到，kfifo 在修改数据和更新指针（队头，队尾）之间，通过函数 smp_wmb() 插入了一个 Store Barrier，从而确保了：

- 更新指针的操作，不会被重排序到修改数据之前。
- 更新指针的时候，Store Cache 被刷新，其他 CPU 可见。

这里有一个关键点要说明：在修改 in、out 指针之后，并没有插入内存屏障。这意味着对 in、out 的修改，可能短时间内会对其他 CPU 或线程不可见。也就是说，数据修改了，但是指针没变。但这不会引发问题，最多是在读的时候队列不为空，判断为空了；在写的时候，队列没满，但判断为满了。调用者的重试机制解决了这个问题。因此，这里其实是"弱一致的"，或者说是"最终一致性"的。

1.7.2　JDK 中的内存屏障

内存屏障是很底层的概念，对于 Java 开发者来说，一般用 volatile 关键字就足够了。但从 JDK 8 开始，Java 在 Unsafe 类中提供了三个内存屏障函数，如下所示。

```
public final class Unsafe {
…
public native void loadFence();
public native void storeFence();
public native void fullFence();
…
}
```

要说明的是，这三个屏障并不是最基本的内存屏障。在理论层面，可以把基本的 CPU 内存屏障分成四种：

（1）LoadLoad：禁止读和读的重排序。

（2）StoreStore：禁止写和写的重排序。

（3）LoadStore：禁止读和写的重排序。

（4）StoreLoad：禁止写和读的重排序。

JDK 定义的三种内存屏障和理论层面划分的四类内存屏障是什么关系呢？

在 JDK 8 的代码注释中，只有零星的一两句话。但 JDK 9 对这个问题做了详尽的解释，如下所示。

```
/**
 + * Ensures that loads before the fence will not be reordered with loads and
 + * stores after the fence; a "LoadLoad plus LoadStore barrier".
 + *
 + * Corresponds to C11 atomic_thread_fence(memory_order_acquire)
 + * (an "acquire fence").
 + *
 + * A pure LoadLoad fence is not provided, since the addition of LoadStore
 + * is almost always desired, and most current hardware instructions that
```

```
+     * provide a LoadLoad barrier also provide a LoadStore barrier for free.
      * @since 1.8
      */
     public native void loadFence();
     /**
+     * Ensures that loads and stores before the fence will not be reordered with
+     * stores after the fence; a "StoreStore plus LoadStore barrier".
+     *
+     * Corresponds to C11 atomic_thread_fence(memory_order_release)
+     * (a "release fence").
+     *
+     * A pure StoreStore fence is not provided, since the addition of LoadStore
+     * is almost always desired, and most current hardware instructions that
+     * provide a StoreStore barrier also provide a LoadStore barrier for free.
      * @since 1.8
      */
     public native void storeFence();

    /**
+    * Ensures that loads and stores before the fence will not be reordered
+    * with loads and stores after the fence. Implies the effects of both
+    * loadFence() and storeFence(), and in addition, the effect of a StoreLoad
+    * barrier.
+    *
+    * Corresponds to C11 atomic_thread_fence(memory_order_seq_cst).
     * @since 1.8
     */
    public native void fullFence();
```

根据注释，我们知道：

loadFence = LoadLoad + LoadStore

storeFence = StoreStore + LoadStore

fullFence = loadFence + storeFence + StoreLoad

1.7.3　volatile 实现原理

由于不同的 CPU 架构的缓存体系不一样，重排序的策略不一样，所提供的内存屏障指令也就有差异。

这里只探讨为了实现 volatile 关键字的语义的一种参考做法：

（1）在 volatile 写操作的前面插入一个 StoreStore 屏障。保证 volatile 写操作不会和之前的写操作重排序。

（2）在 volatile 写操作的后面插入一个 StoreLoad 屏障。保证 volatile 写操作不会和之后的读操作重排序。

（3）在 volatile 读操作的后面插入一个 LoadLoad 屏障 + LoadStore 屏障。保证 volatile 读操作不会和之后的读操作、写操作重排序。

具体到 x86 平台上，其实不会有 LoadLoad、LoadStore 和 StoreStore 重排序，只有 StoreLoad 一种重排序（内存屏障），也就是只需要在 volatile 写操作后面加上 StoreLoad 屏障。其他的 CPU 架构做法不一，此处就不再进一步深入讨论。

1.8 final 关键字

1.8.1 构造函数溢出问题

考虑下面的代码：

```
public class Example{
   private int i;
   private int j;
   private static Example obj;
   public Example()   {
      i=1;
      j=2;
   }
   public static void write()      //线程A先执行write()
   {
     obj = new Example()
   }
   public static void read()       //线程B再执行read()
   {
     if(obj!=null)
     {
        int a = obj.i;
        int b = obj.j;           //请问，a, b是否一定等于1, 2？
     }
   }
}
```

答案是：a，b 未必一定等于 1，2。和 DCL 的例子类似，也就是构造函数溢出问题。obj = new Example()这行代码，分解成三个操作：

① 分配一块内存；

② 在内存上初始化 i = 1，j = 2；

③ 把 obj 指向这块内存。

操作②和操作③可能重排序，因此线程 B 可能看到未正确初始化的值。对于构造函数溢出，通俗来讲，就是一个对象的构造并不是"原子的"，当一个线程正在构造对象时，另外一个线程却可以读到未构造好的"一半对象"。

1.8.2 final 的 happen-before 语义

要解决这个问题，不止有一种办法。

办法 1：给 i，j 都加上 volatile 关键字。

办法 2：为 read/write 函数都加上 synchronized 关键字。

如果 i，j 只需要初始化一次，则后续值就不会再变了，还有办法 3，为其加上 final 关键字。之所以能解决问题，是因为同 volatile 一样，final 关键字也有相应的 happen-before 语义：

（1）对 final 域的写（构造函数内部），happen-before 于后续对 final 域所在对象的读。

（2）对 final 域所在对象的读，happen-before 于后续对 final 域的读。

通过这种 happen-before 语义的限定，保证了 final 域的赋值，一定在构造函数之前完成，不会出现另外一个线程读取到了对象，但对象里面的变量却还没有初始化的情形，避免出现构造函数溢出的问题。

关于 final 和 volatile 的特性与背后的原理，到此为止就讲完了，在后续 Concurrent 包的源码分析中会反复看到这两个关键字的身影。接下来总结常用的几个 happen-before 规则。

1.8.3 happen-before 规则总结

（1）单线程中的每个操作，happen-before 于该线程中任意后续操作。

（2）对 volatile 变量的写，happen-before 于后续对这个变量的读。

（3）对 synchronized 的解锁，happen-before 于后续对这个锁的加锁。

（4）对 final 变量的写，happen-before 于 final 域对象的读，happen-before 于后续对 final 变

量的读。

四个基本规则再加上 happen-before 的传递性，就构成 JMM 对开发者的整个承诺。在这个承诺以外的部分，程序都可能被重排序，都需要开发者小心地处理内存可见性问题。

图 1-9 表示了 volatile 背后的原理。

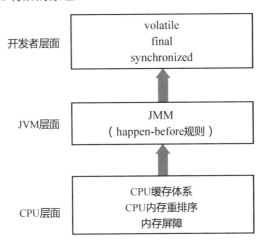

图 1-9　从底向上看 volatile 背后的原理

1.9　综合应用：无锁编程

提到多线程编程，就绕不开"锁"，在 Java 中就是指 synchronized 关键字和 Lock。在 Linux 中，主要是指 pthread 的 mutex。但锁又是性能杀手，所以很多的前辈大师们研究如何可以不用锁，也能实现线程安全。无锁编程是一个庞大而深入的话题，既涉及底层的 CPU 架构（例如前面讲的内存屏障），又涉及不同语言的具体实现。在作者的另一本书《软件架构设计：大型网站技术架构与业务架构融合之道》中，也对无锁编程做了介绍，此处再次总结一下常用的几种无锁编程的场景，也是对本章技术点的一个应用。

1.9.1　一写一读的无锁队列：内存屏障

一写一读的无锁队列即 Linux 内核的 kfifo 队列，一写一读两个线程，不需要锁，只需要内存屏障。

1.9.2　一写多读的无锁队列：volatile 关键字

在 Martin Fowler 关于 LMAX 架构的介绍中，谈到了 Disruptor。Disruptor 是一个开源的并

发框架，能够在无锁的情况下实现 Queue 并发操作。

Disruptor 的 RingBuffer 之所以可以做到完全无锁，也是因为"单线程写"，这是"前提的前提"。离开了这个前提条件，没有任何技术可以做到完全无锁。借用 Disruptor 官方提到的一篇博客文章 *Sharing Data Among Threads Without Contention*，也就是 single-writer principle。

在这个原则下，利用 volatile 关键字可以实现一写多读的线程安全。具体来说，就是 RingBuffer 有一个头指针，对应一个生产者线程；多个尾指针对应多个消费者线程。每个消费者线程只会操作自己的尾指针。所有这些指针的类型都是 volatile 变量，通过头指针和尾指针的比较，判断队列是否为空。

1.9.3 多写多读的无锁队列：CAS

同内存屏障一样，CAS（Compare And Set）也是 CPU 提供的一种原子指令。在第 2 章中会对 CAS 进行详细的解释。

基于 CAS 和链表，可以实现一个多写多读的队列。具体来说，就是链表有一个头指针 head 和尾指针 tail。入队列，通过对 tail 进行 CAS 操作完成；出队列，对 head 进行 CAS 操作完成。

在第 3 章讲 Lock 的实现的时候，将反复用到这种队列，会详细展开介绍。

1.9.4 无锁栈

无锁栈比无锁队列的实现更简单，只需要对 head 指针进行 CAS 操纵，就能实现多线程的入栈和出栈。

在第 4 章讲工具类的实现的时候，会用到无锁栈。

1.9.5 无锁链表

相比无锁队列与无锁栈，无锁链表要复杂得多，因为无锁链表要在中间插入和删除元素。

在 5.6 节，介绍 ConcurrentSkipListMap 实现的时候，会讲到并发的跳查表。其实现就是基于无锁链表的，到时会详细展开论述。

第 2 章 Atomic 类

从本章开始,我们将从简单到复杂,从底层到上层,一步步剖析整个 Concurrent 包的层次体系,如图 2-1 所示。

图 2-1 整个 Concurrent 包的层次体系

2.1 AtomicInteger 和 AtomicLong

如下面代码所示,对于一个整数的加减操作,要保证线程安全,需要加锁,也就是加 synchronized

关键字。

```java
public class Example{
  private int count = 0;
  public void synchronized increment()     {    //线程A调用
    count++
  }
  public void  synchronized decrement()    {    //线程B调用
    count--
  }
}
```

但有了 Concurrent 包的 Atomic 相关的类之后，synchronized 关键字可以用 AtomicInteger 代替，其性能更好，对应的代码变为如下所示。

```java
public class Example{
  private AtomicInteger count= new AtomicInteger(0);
  public void add()    {
    count.getAndIncrement()
  }
  public long decr()    {
    count.getAndDecrement()
  }
}
```

其对应的源码如下（源自 JDK 7）：

```java
public class AtomicInteger extends Number implements java.io.Serializable {
…
private volatile int value;        //封装了一个int变量，对其进行CAS操作
…
public final int getAndIncrement() {
for (;;) {
    int current = get();
    int next = current + 1;
    if (compareAndSet(current, next))
        return current;
}
}
public final int getAndDecrement() {
for (;;) {
    int current = get();
    int next = current - 1;
    if (compareAndSet(current, next))
        return current;
```

            }
    }
}
其源码很简单，但却反映了几个很重要的思想，下面一一说明。

2.1.1 悲观锁与乐观锁

对于悲观锁，作者认为数据发生并发冲突的概率很大，所以读操作之前就上锁。synchronized 关键字，以及后面要讲的 ReentrantLock 都是悲观锁的典型例子。

对于乐观锁，作者认为数据发生并发冲突的概率比较小，所以读操作之前不上锁。等到写操作的时候，再判断数据在此期间是否被其他线程修改了。如果被其他线程修改了，就把数据重新读出来，重复该过程；如果没有被修改，就写回去。判断数据是否被修改，同时写回新值，这两个操作要合成一个原子操作，也就是 CAS（Compare And Set）。

AtomicInteger 的实现就是典型的乐观锁，在 MySQL 和 Redis 中有类似的思路。

2.1.2 Unsafe 的 CAS 详解

上面调用的 CAS 函数，其实是封装的 Unsafe 类中的一个 native 函数，如下所示。

```
public final boolean compareAndSet(int expect, int update) {
    return unsafe.compareAndSwapInt(this, valueOffset, expect, update);
}
```

AtomicInteger 封装过的 compareAndSet 有两个参数。第一个参数 expect 是指变量的旧值（是读出来的值，写回去的时候，希望没有被其他线程修改，所以称为 expect）；第二个参数 update 是指变量的新值（修改过的，希望写入的值）。当 expect 等于变量当前的值时，说明在修改的期间，没有其他线程对此变量进行过修改，所以可以成功写入，变量被更新为 update，返回 true；否则返回 false。

Unsafe 类是整个 Concurrent 包的基础，里面所有函数都是 native 的。具体到 compareAndSwapInt 函数，如下所示。

```
public final native boolean compareAndSwapInt(Object var1, long var2, int var4, int var5);
```

该函数有 4 个参数。在前两个参数中，第一个是对象（也就是 AtomicInteger 对象），第二个是对象的成员变量（也就是 AtomictInteger 里面包的 int 变量 value），后两个参数保持不变。

要特别说明一下第二个参数，它是一个 long 型的整数，经常被称为 xxxOffset，意思是某个成员变量在对应的类中的内存偏移量（该变量在内存中的位置），表示该成员变量本身。在 Unsafe

中专门有一个函数,把成员变量转化成偏移量,如下所示。

```
public native long objectFieldOffset(Field var1);
```

所有调用 CAS 的地方,都会先通过这个函数把成员变量转换成一个 Offset。以 AtomicInteger 为例:

```
public class AtomicInteger extends Number implements java.io.Serializable {
...
  private static final Unsafe unsafe = Unsafe.getUnsafe();
private static final long valueOffset;

static {
  try {
    valueOffset = unsafe.objectFieldOffset
        (AtomicInteger.class.getDeclaredField("value"));
  } catch (Exception ex) { throw new Error(ex); }
}
...
}
```

从上面代码可以看到,无论是 Unsafe 还是 valueOffset,都是静态的,也就是类级别的,所有对象共用的。

在转化的时候,先通过反射(getDeclaredField)获取 value 成员变量对应的 Field 对象,再通过 objectFieldOffset 函数转化成 valueOffset。此处的 valueOffset 就代表了 value 变量本身,后面执行 CAS 操作的时候,不是直接操作 value,而是操作 valueOffset。

2.1.3 自旋与阻塞

当一个线程拿不到锁的时候,有以下两种基本的等待策略。

策略 1:放弃 CPU,进入阻塞状态,等待后续被唤醒,再重新被操作系统调度。

策略 2:不放弃 CPU,空转,不断重试,也就是所谓的"自旋"。

很显然,如果是单核的 CPU,只能用策略 1。因为如果不放弃 CPU,那么其他线程无法运行,也就无法释放锁。但对于多 CPU 或者多核,策略 2 就很有用了,因为没有线程切换的开销。

AtomicInteger 的实现就用的是"自旋"策略,如果拿不到锁,就会一直重试。

有一点要说明:这两种策略并不是互斥的,可以结合使用。如果拿不到锁,先自旋几圈;如果自旋还拿不到锁,再阻塞,synchronized 关键字就是这样的实现策略。

除了 AtomicInteger,AtomicLong 也是同样的原理,此处不再赘述。

2.2　AtomicBoolean 和 AtomicReference

2.2.1　为什么需要 AtomicBoolean

对于 int 或者 long 型变量，需要进行加减操作，所以要加锁；但对于一个 boolean 类型来说，true 或 false 的赋值和取值操作，加上 volatile 关键字就够了，为什么还需要 AtomicBoolean 呢？

这是因为往往要实现下面这种功能：

```
if(flag == false) {
  flag = true;
  ...
}
```

也就是要实现 compare 和 set 两个操作合在一起的原子性，而这也正是 CAS 提供的功能。上面的代码，就变成：

```
if(compareAndSet(false,true)) {
...
}
```

同样地，AtomicReference 也需要同样的功能，对应的函数如下：

```
public final boolean compareAndSet(V expect, V update) {
    return unsafe.compareAndSwapObject(this, valueOffset, expect, update);
}
```

其中，expect 是旧的引用，update 为新的引用。

2.2.2　如何支持 boolean 和 double 类型

在 Unsafe 类中，只提供了三种类型的 CAS 操作：int、long、Object（也就是引用类型）。如下所示。

```
public final class Unsafe {
...
public final native boolean compareAndSwapObject(Object var1, long var2, Object var4, Object var5);
public final native boolean compareAndSwapInt(Object var1, long var2, int var4, int var5);
public final native boolean compareAndSwapLong(Object var1, long var2, long var4, long var6);
...
}
```

第一个参数是要修改的对象，第二个参数是对象的成员变量在内存中的位置（一个 long 型的整数），第三个参数是该变量的旧值，第四个参数是该变量的新值。

AtomicBoolean 类型怎么支持呢？

对于用 int 型来代替的，在入参的时候，将 boolean 类型转换成 int 类型；在返回值的时候，将 int 类型转换成 boolean 类型。如下所示。

```java
public class AtomicBoolean implements java.io.Serializable {
…
private volatile int value;
public final boolean get() {
return value != 0;
}
public final boolean compareAndSet(boolean expect, boolean update) {
int e = expect ? 1 : 0;
int u = update ? 1 : 0;
return unsafe.compareAndSwapInt(this, valueOffset, e, u);
}
…
}
```

如果是 double 类型，又如何支持呢？

这依赖 double 类型提供的一对 double 类型和 long 类型互转的函数，这点在介绍 DoubleAdder 的时候会提到。

```java
public final class Double extends Number implements Comparable<Double> {
…
public static native double longBitsToDouble(long bits);
public static native long doubleToRawLongBits(double value);
…
}
```

2.3　AtomicStampedReference 和 AtomicMarkableReference

2.3.1　ABA 问题与解决办法

到目前为止，CAS 都是基于"值"来做比较的。但如果另外一个线程把变量的值从 A 改为 B，再从 B 改回到 A，那么尽管修改过两次，可是在当前线程做 CAS 操作的时候，却会因为值没变而认为数据没有被其他线程修改过，这就是所谓的 ABA 问题。

要解决 ABA 问题，不仅要比较"值"，还要比较"版本号"，而这正是 AtomicStampedReference 做的事情，其对应的 CAS 函数如下：

```
public boolean compareAndSet(V    expectedReference,
                V   newReference,
                    int expectedStamp,
                    int newStamp)
```

之前的 CAS 只有两个参数，这里的 CAS 有四个参数，后两个参数就是版本号的旧值和新值。

当 expectedReference != 对象当前的 reference 时，说明该数据肯定被其他线程修改过；

当 expectedReference == 对象当前的 reference 时，再进一步比较 expectedStamp 是否等于对象当前的版本号，以此判断数据是否被其他线程修改过。

2.3.2 为什么没有 AtomicStampedInteger 或 AtomictStampedLong

要解决 Integer 或者 Long 型变量的 ABA 问题，为什么只有 AtomicStampedReference，而没有 AtomicStampedInteger 或者 AtomictStampedLong 呢？

因为这里要同时比较数据的"值"和"版本号"，而 Integer 型或者 Long 型的 CAS 没有办法同时比较两个变量，于是只能把值和版本号封装成一个对象，也就是这里面的 Pair 内部类，然后通过对象引用的 CAS 来实现。代码如下所示。

```
public class AtomicStampedReference<V> {
private static class Pair<T> {
   final T reference;
   final int stamp;
   private Pair(T reference, int stamp) {
      this.reference = reference;
      this.stamp = stamp;
   }
   static <T> Pair<T> of(T reference, int stamp) {
      return new Pair<T>(reference, stamp);
   }
}
}

private volatile Pair<V> pair;
   public boolean compareAndSet(V    expectedReference,
                   V   newReference,
                       int expectedStamp,
                       int newStamp) {
```

```
        Pair<V> current = pair;
        return
            expectedReference == current.reference &&
            expectedStamp == current.stamp &&
            ((newReference == current.reference &&
              newStamp == current.stamp) ||
             casPair(current, Pair.of(newReference, newStamp)));
    }
    private boolean casPair(Pair<V> cmp, Pair<V> val) {
        return UNSAFE.compareAndSwapObject(this, pairOffset, cmp, val);
    }
    ...
}
```

当使用的时候，在构造函数里面传入值和版本号两个参数，应用程序对版本号进行累加操作，然后调用上面的 CAS。如下所示。

```
    public AtomicStampedReference(V initialRef, int initialStamp) {
        pair = Pair.of(initialRef, initialStamp);
    }
```

2.3.3　AtomicMarkableReference

AtomicMarkableReference 与 AtomicStampedReference 原理类似，只是 Pair 里面的版本号是 boolean 类型的，而不是整型的累加变量，如下所示。

```
public class AtomicMarkableReference<V> {
    private static class Pair<T> {
        final T reference;
        final boolean mark;    //boolean 类型的版本号
        private Pair(T reference, boolean mark) {
            this.reference = reference;
            this.mark = mark;
        }
        static <T> Pair<T> of(T reference, boolean mark) {
            return new Pair<T>(reference, mark);
        }
    }
    private volatile Pair<V> pair;
    ...
}
```

因为是 boolean 类型，只能有 true、false 两个版本号，所以并不能完全避免 ABA 问题，只是降低了 ABA 发生的概率。

2.4 AtomicIntegerFieldUpdater、AtomicLongFieldUpdater 和 AtomicReferenceFieldUpdater

2.4.1 为什么需要 AtomicXXXFieldUpdater

如果一个类是自己编写的，则可以在编写的时候把成员变量定义为 Atomic 类型。但如果是一个已经有的类，在不能更改其源代码的情况下，要想实现对其成员变量的原子操作，就需要 AtomicIntegerFieldUpdater、AtomicLongFieldUpdater 和 AtomicReferenceFieldUpdater。下面以 AtomicIntegerFieldUpdater 为例介绍其实现原理。

首先，其构造函数是 protected，不能直接构造其对象，必须通过它提供的一个静态函数来创建，如下所示。

```
    protected AtomicIntegerFieldUpdater() {
}
    public static <U> AtomicIntegerFieldUpdater<U> newUpdater(Class<U> tclass,
String fieldName) {
    ...
}
```

newUpdater(..)静态函数传入的是要修改的类（不是对象）和对应的成员变量的名字，内部通过反射拿到这个类的成员变量，然后包装成一个 AtomicIntegerFieldUpdater 对象。所以，这个对象表示的是类的某个成员，而不是对象的成员变量。

若要修改某个对象的成员变量的值，再传入相应的对象，如下所示。

```
    public int getAndIncrement(T obj) {
    int prev, next;
    do {
        prev = get(obj);
        next = prev + 1;
    } while (!compareAndSet(obj, prev, next));
    return prev;
}
    public final boolean compareAndSet(T obj, int expect, int update) {
    accessCheck(obj);
    return U.compareAndSwapInt(obj, offset, expect, update);
}
```

accecssCheck 函数的作用是检查该 obj 是不是 tclass 类型，如果不是，则拒绝修改，抛出异常。

从代码可以看到，其 CAS 原理和 AtomictInteger 是一样的，底层都调用了 Unsafe 的 compareAndSwapInt(..)函数。

2.4.2 限制条件

要想使用 AtomicIntegerFieldUpdater 修改成员变量，成员变量必须是 volatile 的 int 类型（不能是 Integer 包装类），该限制从其构造函数中可以看到：

```java
    public static <U> AtomicIntegerFieldUpdater<U> newUpdater(Class<U> tclass,
                                                String fieldName) {
return new AtomicIntegerFieldUpdaterImpl<U>
        (tclass, fieldName, Reflection.getCallerClass());
}
    AtomicIntegerFieldUpdaterImpl(final Class<T> tclass, final String fieldName,
final Class<?> caller){
    ...
    if (field.getType() != int.class)
        throw new IllegalArgumentException("Must be integer type");

if (!Modifier.isVolatile(modifiers))
    throw new IllegalArgumentException("Must be volatile type");
    ...
    }
```

至于 AtomicLongFieldUpdater、AtomicReferenceFieldUpdater，也有类似的限制条件。其底层的 CAS 原理，也和 AtomicLong、AtomicReference 一样，此处不再赘述。

2.5 AtomicIntegerArray、AtomicLongArray 和 AtomicReferenceArray

Concurrent 包提供了 AtomicIntegerArray、AtomicLongArray、AtomicReferenceArray 三个数组元素的原子操作。注意，这里并不是说对整个数组的操作是原子的，而是针对数组中一个元素的原子操作而言。

2.5.1 使用方式

以 AtomicIntegerArray 为例，其使用方式如下：

```java
    public final int getAndIncrement(int i) {
    return getAndAdd(i, 1);
}
```

相比于 AtomicInteger 的 getAndIncrement()函数，这里只是多了一个传入参数：数组的下标 i。

其他函数也与此类似，相比于 AtomicInteger 的各种加减函数，也都是多一个下标 i，如下所示。

```
public final boolean compareAndSet(int i, int expect, int update) {…}
public final int getAndDecrement(int i) {…}
public final int getAndSet(int i, int newValue) {…}
```

2.5.2 实现原理

其底层的 CAS 函数用的还是 compareAndSwapInt，但是把数组下标 i 转化成对应的内存偏移量，所用的方法和之前的 AtomicInteger 不太一样，如下所示。

```
private static long byteOffset(int i) {
return ((long) i << shift) + base;
}
```

把下标 i 转换成对应的内存地址，用到了 shift 和 base 两个变量。这两个变量都是 AtomicIntegerArray 的静态成员变量，用了 Unsafe 类的 arrayBaseOffset 和 arrayIndexScale 两个函数来获取。赋值如下：

```
private static final int base = unsafe.arrayBaseOffset(int[].class);
static {
int scale = unsafe.arrayIndexScale(int[].class);   //scale 刚好是 2 的整数次方
if ((scale & (scale - 1)) != 0)
    throw new Error("data type scale not a power of two");
shift = 31 - Integer.numberOfLeadingZeros(scale);
}
```

其中，base 表示数组的首地址的位置，scale 表示一个数组元素的大小，i 的偏移量则等于：i * scale + base。

但为了优化性能，使用了位移操作，shift 表示 scale 中 1 的位置（scale 是 2 的整数次方）。所以，偏移量的计算变成上面代码中的：i << shift + base，表达的意思就是：i * scale + base。

知道了偏移量的计算方式，理解 CAS 操作就容易了：

```
public final int getAndAdd(int i, int delta) {
long offset = checkedByteOffset(i);        //下标 i 转换成数组中的内存偏移量（也就是
                                           //上面的计算公式）
while (true) {
    int current = getRaw(offset);          //根据偏移量获取下标 i 的旧值
    if (compareAndSetRaw(offset, current, current + delta))
        return current;
```

```
    }
}
    private boolean compareAndSetRaw(long offset, int expect, int update) {
        return unsafe.compareAndSwapInt(array, offset, expect, update);
}
```

第 1 个参数是 int[]对象，第 2 个参数是下标 i 对应的内存偏移量，第 3 个和第 4 个参数分别是旧值和新值。

明白了 AtomicIntegerArray 的实现原理，另外两个数组的原子类实现原理与之类似，此处不再赘述。

2.6　Striped64 与 LongAdder

从 JDK 8 开始，针对 Long 型的原子操作，Java 又提供了 LongAdder、LongAccumulator；针对 Double 类型，Java 提供了 DoubleAdder、DoubleAccumulator。Striped64 相关的类的继承层次如图 2-2 所示。

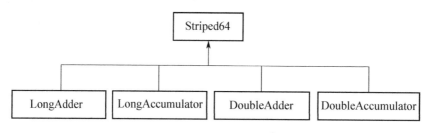

图 2-2　Striped64 相关的类的继承层次

读者会有一个疑问，既然已有了 AtomicLong，为什么还要提供 LongAdder？并且还提供了 LongAccumulator？

2.6.1　LongAdder 原理

AtomicLong 内部是一个 volatile long 型变量，由多个线程对这个变量进行 CAS 操作。多个线程同时对一个变量进行 CAS 操作，在高并发的场景下仍不够快，如果再要提高性能，该怎么做呢？

把一个变量拆成多份，变为多个变量，有些类似于 ConcurrentHashMap 的分段锁的例子。如图 2-3 所示，把一个 Long 型拆成一个 base 变量外加多个 Cell，每个 Cell 包装了一个 Long 型变量。当多个线程并发累加的时候，如果并发度低，就直接加到 base 变量上；如果并发度高，冲突大，平摊到这些 Cell 上。在最后取值的时候，再把 base 和这些 Cell 求 sum 运算。

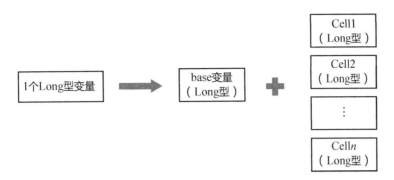

图 2-3　1 个 Long 型变量被拆成多个 Long 型的示意图

以 LongAdder 的 sum() 函数为例,如下所示。

```
public long sum() {
Cell[] as = cells; Cell a;
long sum = base;
if (as != null) {
    for (int i = 0; i < as.length; ++i) {
        if ((a = as[i]) != null)
            sum += a.value;
    }
}
return sum;
}
```

由于无论是 long,还是 double,都是 64 位的。但因为没有 double 型的 CAS 操作,所以是通过把 double 型转化成 long 型来实现的。所以,上面的 base 和 cell[] 变量,是位于基类 Striped64 当中的。英文 Striped 意为"条带",也就是分片。

```
abstract class Striped64 extends Number {
transient volatile long base;
transient volatile Cell[] cells;
@sun.misc.Contended static final class Cell {
volatile long value;
…}
}
```

2.6.2　最终一致性

在 sum 求和函数中,并没有对 cells[] 数组加锁。也就是说,一边有线程对其执行求和操作,一边还有线程修改数组里的值,也就是最终一致性,而不是强一致性。这也类似于

ConcurrentHashMap 中的 clear() 函数，一边执行清空操作，一边还有线程放入数据，clear() 函数调用完毕后再读取，hash map 里面可能还有元素。因此，在 LongAdder 开篇的注释中，把它和 AtomicLong 的使用场景做了比较。它适合高并发的统计场景，而不适合要对某个 Long 型变量进行严格同步的场景。

2.6.3 伪共享与缓存行填充

在 Cell 类的定义中，用了一个独特的注解 @sun.misc.Contended，这是 JDK 8 之后才有的，背后涉及一个很重要的优化原理：伪共享与缓存行填充。

在讲 CPU 架构的时候提到过，每个 CPU 都有自己的缓存。缓存与主内存进行数据交换的基本单位叫 Cache Line（缓存行）。在 64 位 x86 架构中，缓存行是 64 字节，也就是 8 个 Long 型的大小。这也意味着当缓存失效，要刷新到主内存的时候，最少要刷新 64 字节。

如图 2-4 所示，主内存中有变量 X、Y、Z（假设每个变量都是一个 Long 型），被 CPU1 和 CPU2 分别读入自己的缓存，放在了同一行 Cache Line 里面。当 CPU1 修改了 X 变量，它要失效整行 Cache Line，也就是往总线上发消息，通知 CPU 2 对应的 Cache Line 失效。由于 Cache Line 是数据交换的基本单位，无法只失效 X，要失效就会失效整行的 Cache Line，这会导致 Y、Z 变量的缓存也失效。

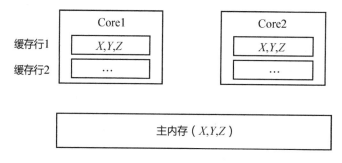

图 2-4　伪共享示意图

虽然只修改了 X 变量，本应该只失效 X 变量的缓存，但 Y、Z 变量也随之失效。Y、Z 变量的数据没有修改，本应该很好地被 CPU1 和 CPU2 共享，却没做到，这就是所谓的"伪共享问题"。

问题的原因是，Y、Z 和 X 变量处在了同一行 Cache Line 里面。要解决这个问题，需要用到所谓的"缓存行填充"，分别在 X、Y、Z 后面加上 7 个无用的 Long 型，填充整个缓存行，让 X、Y、Z 处在三行不同的缓存行中，如图 2-5 所示。

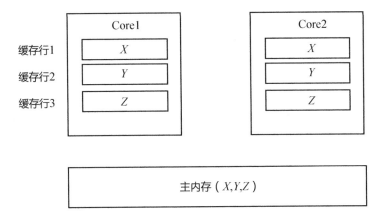

图 2-5 缓存行填充示意图

下面的代码来自 JDK 7 的 Exchanger 类，为了安全，它填充了 15（8+7）个 long 型。

```
private static final class Slot extends AtomicReference<Object> {
long q0, q1, q2, q3, q4, q5, q6, q7, q8, q9, qa, qb, qc, qd, qe;
}
```

在著名的开源无锁并发框架 Disruptor 中，也有类似的代码：

```
abstract class RingBufferPad{
    protected long p1, p2, p3, p4, p5, p6, p7;  //填充了 7 个 long 型数据
}
abstract class RingBufferFields<E> extends RingBufferPad{
...
}
```

而在 JDK 8 中，就不需要写这种晦涩的代码了，只需声明一个@sun.misc.Contended 即可。下面的代码摘自 JDK 8 里面 Exchanger 中 Node 的定义，相较于 JDK 7 有了明显变化。

```
@sun.misc.Contended static final class Node {
...
}
```

回到上面的例子，之所以这个地方要用缓存行填充，是为了不让 Cell[]数组中相邻的元素落到同一个缓存行里。

2.6.4 LongAdder 核心实现

下面来看 LongAdder 最核心的累加函数 add(long x)，自增、自减操作都是通过调用该函数实现的。

```
public void increment() { add(1); }
```

```java
public void decrement() { add(-1); }
public void add(long x) {
Cell[] as; long b, v; int m; Cell a;
if ((as = cells) != null || !casBase(b = base, b + x)) {     //第1次尝试
    boolean uncontended = true;
    if (as == null || (m = as.length - 1) < 0 ||
        (a = as[getProbe() & m]) == null ||
        !(uncontended = a.cas(v = a.value, v + x)))          //第2次尝试
        longAccumulate(x, null, uncontended);
    }
}
```

当一个线程调用 add(x) 的时候，首先会尝试使用 casBase 把 x 加到 base 变量上。如果不成功，则再用 a.cas(..) 函数尝试把 x 加到 Cell 数组的某个元素上。如果还不成功，最后再调用 longAccumulate(..) 函数。

> **注意**：Cell[] 数组的大小始终是 2 的整数次方，在运行中会不断扩容，每次扩容都是增长 2 倍。上面代码中的 as[getProbe() & m] 其实就是对数组的大小取模。因为 m = as.length–1，getProbe() 为该线程生成一个随机数，用该随机数对数组的长度取模。因为数组长度是 2 的整数次方，所以可以用 & 操作来优化取模运算。

对于一个线程来说，它并不在意到底是把 x 累加到 base 上面，还是累加到 Cell[] 数组上面，只要累加成功就可以。因此，这里使用随机数来实现 Cell 的长度取模。

如果两次尝试都不成功，则调用 longAccumulate(..) 函数，该函数在 Striped64 里面 LongAccumulator 也会用到，如下所示。

```java
final void longAccumulate(long x, LongBinaryOperator fn, boolean wasUncontended) {
    int h;
    if ((h = getProbe()) == 0) {
        ThreadLocalRandom.current();
        h = getProbe();
        wasUncontended = true;
    }
    boolean collide = false;           // 写入冲突的标志位
    for (;;) {
        Cell[] as; Cell a; int n; long v;
        if ((as = cells) != null && (n = as.length) > 0) {
            if ((a = as[(n - 1) & h]) == null) {
                if (cellsBusy == 0) {
                    Cell r = new Cell(x);
```

```java
            if (cellsBusy == 0 && casCellsBusy()) {
                boolean created = false;
                try {
                    Cell[] rs; int m, j;
                    if ((rs = cells) != null &&
                        (m = rs.length) > 0 &&
                        rs[j = (m - 1) & h] == null) {
                        rs[j] = r;
                        created = true;
                    }
                } finally {
                    cellsBusy = 0;
                }
                if (created)
                    break;
                continue;
            }
        }
        collide = false;
    }
    else if (!wasUncontended)
        wasUncontended = true;
    else if (a.cas(v = a.value, ((fn == null) ? v + x : fn.applyAsLong(v, x))))
        break;
    else if (n >= NCPU || cells != as)
        collide = false;
    else if (!collide)
        collide = true;
    else if (cellsBusy == 0 && casCellsBusy()) {
        try {
            if (cells == as) {          // 对数组扩容
                Cell[] rs = new Cell[n << 1];
                for (int i = 0; i < n; ++i)
                    rs[i] = as[i];
                cells = rs;
            }
        } finally {
            cellsBusy = 0;
        }
        collide = false;
        continue;
    }
```

```
            h = advanceProbe(h);
        }
        else if (cellsBusy == 0 && cells == as && casCellsBusy()) {
            boolean init = false;
            try {                               // 初始化数组
                if (cells == as) {
                    Cell[] rs = new Cell[2];
                    rs[h & 1] = new Cell(x);
                    cells = rs;
                    init = true;
                }
            } finally {
                cellsBusy = 0;
            }
            if (init)
                break;
        }
        else if (casBase(v = base, ((fn == null) ? v + x :
                                    fn.applyAsLong(v, x))))
            break;
    }
}
```

上面的函数 fn 就是 2.6.5 节 LongAccumulator 要用到的，但对于 LongAdder 而言，fn = null，就是简单的累加操作 v+x。

上面的 for 循环被分成三个大的分支。在第二个分支里面，进行了 Cells[]数组的初始化工作，初始大小为 2，然后把 x 累加在 0 下标或者 1 下标对应的 Cell 上面。

```
if (cells == as) {
    Cell[] rs = new Cell[2];
    rs[h & 1] = new Cell(x);
    cells = rs;
    init = true;
}
```

在第一个大的分支里面，完成 Cells[]数组的不断扩容，每次扩容都是增长 2 倍。

```
if (cells == as) {
    Cell[] rs = new Cell[n << 1];           //扩容
    for (int i = 0; i < n; ++i)             //拷贝旧数据
        rs[i] = as[i];
    cells = rs;
}
```

数组为空,并且有一个线程正在进行初始化工作,于是进入第三个大的分支中,尝试对 base 变量进行累积,如果再次失败,则会再次进入第一个大的分支。

```
else if (casBase(v = base, ((fn == null) ? v + x : fn.applyAsLong(v, x))))
break;
```

2.6.5 LongAccumulator

LongAccumulator 的原理和 LongAdder 类似,只是功能更强大,下面为两者构造函数的对比:

```
public LongAdder() {}
public LongAccumulator(LongBinaryOperator accumulatorFunction, long identity)
{}
```

LongAdder 只能进行累加操作,并且初始值默认为 0;LongAccumulator 可以自己定义一个二元操作符,并且可以传入一个初始值。

```
public interface LongBinaryOperator {
long applyAsLong(long left, long right);
}
```

操作符的左值,就是 base 变量或者 Cells[]中元素的当前值;右值,就是 add()函数传入的参数 x。

下面是 LongAccumulator 的 accumulate(x)函数,与 LongAdder 的 add(x)函数类似,最后都是调用的 Striped64 的 LongAccumulate(..)函数。唯一的差别就是 LongAdder 的 add(x)函数调用的是 casBase(b, b+x),这里调用的是 casBase(b,r),其中,r = function.applyAsLong(b = base, x)。

```
public void accumulate(long x) {
Cell[] as; long b, v, r; int m; Cell a;
if ((as = cells) != null ||
   (r = function.applyAsLong(b = base, x)) != b && !casBase(b, r)) {
   boolean uncontended = true;
   if (as == null || (m = as.length - 1) < 0 ||
      (a = as[getProbe() & m]) == null ||
      !(uncontended =
        (r = function.applyAsLong(v = a.value, x)) == v ||
        a.cas(v, r)))
      longAccumulate(x, function, uncontended);
   }
}
```

2.6.6 DoubleAdder 与 DoubleAccumulator

DoubleAdder 其实也是用 long 型实现的,因为没有 double 类型的 CAS 函数。下面是

DoubleAdder 的 add(x)函数，和 LongAdder 的 add(x)函数基本一样，只是多了 long 和 double 类型的相互转换。

```
public void add(double x) {
Cell[] as; long b, v; int m; Cell a;
if ((as = cells) != null ||
    !casBase(b = base,
            Double.doubleToRawLongBits
            (Double.longBitsToDouble(b) + x))) {
    boolean uncontended = true;
    if (as == null || (m = as.length - 1) < 0 ||
        (a = as[getProbe() & m]) == null ||
        !(uncontended = a.cas(v = a.value,
                          Double.doubleToRawLongBits
                          (Double.longBitsToDouble(v) + x))))
        doubleAccumulate(x, null, uncontended);
    }
}
```

其中的关键 Double.doubleToRawLongBits(Double.longBitsToDouble(b) + x)，在读出来的时候，它把 long 类型转换成 double 类型，然后进行累加，累加的结果再转换成 long 类型，通过 CAS 写回去。

DoubleAccumulate 也是 Striped64 的成员函数，和 longAccumulate 类似，也是多了 long 类型和 double 类型的互相转换。

DoubleAccumulator 和 DoubleAdder 的关系，与 LongAccumulator 和 LongAdder 的关系类似，只是多了一个二元操作符，此处不再赘述。

到此为止，Concurrent 包的所有原子类都介绍完了，接下来分析锁的实现。

第 3 章
Lock 与 Condition

3.1 互斥锁

3.1.1 锁的可重入性

因为在 Concurrent 包中的锁都是"可重入锁",所以一般都命名为 ReentrantX。"可重入锁"是指当一个线程调用 object.lock() 拿到锁,进入互斥区后,再次调用 object.lock(),仍然可以拿到该锁。很显然,通常的锁都要设计成可重入的,否则就会发生死锁。

第 2 章讲的 synchronized 关键字,同样是可重入锁。考虑下面的典型场景:在一个 synchronized 函数 f1() 里面调用另外一个 synchronized 函数 f2()。如果 synchronized 关键字不可重入,那么在 f2() 处就会发生阻塞,这显然不可行。

```
public void synchronized f1() {
…
f2();
}
public void synchronized f2() { …}
```

3.1.2 类的继承层次

在正式介绍锁的实现原理之前,先看一下 Concurrent 包中的与互斥锁(ReentrantLock)相关类之间的继承层次,如图 3-1 所示。

在图 3-1 中,I 表示接口(Interface),A 表示抽象类(Abstract Class),C 表示类(Class),$ 表示内部类。实线表示继承关系,虚线表示引用关系。本书中所有关于类的继承层次的图片都遵循这个图例。

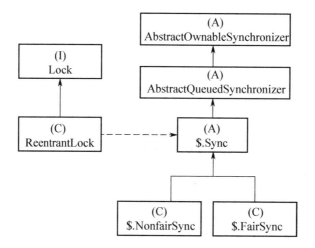

图 3-1 与 ReentrantLock 相关类之间的继承层次

Lock 是一个接口，其定义如下：

```
public interface Lock {
    void lock();
    void lockInterruptibly() throws InterruptedException;
    boolean tryLock();
    boolean tryLock(long time, TimeUnit unit) throws InterruptedException;
    void unlock();
    Condition newCondition();
}
```

常用的方法是 lock()/unlock()。lock() 不能被中断，对应的 lockInterruptibly() 可以被中断。

ReentrantLock 本身没有代码逻辑，实现都在其内部类 Sync 中。

```
public class ReentrantLock implements Lock, java.io.Serializable {
  private final Sync sync;

    public void lock() {
        sync.lock();
    }
    public void unlock() {
        sync.release(1);
    }
    ...
}
```

3.1.3 锁的公平性与非公平性

Sync 是一个抽象类，它有两个子类 FairSync 与 NonfairSync，分别对应公平锁和非公平锁。从下面的 ReentrantLock 构造函数可以看出，会传入一个布尔类型的变量 fair 指定锁是公平的还是非公平的，默认为非公平的。

```
public ReentrantLock() {
    sync = new NonfairSync();
}
public ReentrantLock(boolean fair) {
    sync = fair ? new FairSync() : new NonfairSync();
}
```

什么叫公平锁和非公平锁呢？先举个现实生活中的例子，一个人去火车站售票窗口买票，发现现场有人排队，于是他排在队伍末尾，遵循先到者优先服务的规则，这叫公平；如果他去了不排队，直接冲到窗口买票，这叫不公平。

对应到锁的例子，一个新的线程来了之后，看到有很多线程在排队，自己排到队伍末尾，这叫公平；线程来了之后直接去抢锁，这叫不公平。不同于现实生活，这里默认设置的是非公平锁，其实是为了提高效率，减少线程切换。

后面会详细地通过代码来对比公平锁和非公平锁在实现上的差异。

3.1.4 锁实现的基本原理

Sync 的父类 AbstractQueuedSynchronizer 经常被称作队列同步器（AQS），这个类非常关键，下面会反复提到，该类的父类是 AbstractOwnableSynchronizer。

上一章讲的 Atomic 类都是"自旋"性质的锁，而本章讲的锁将具备 synchronized 功能，也就是可以阻塞一个线程。为了实现一把具有阻塞或唤醒功能的锁，需要几个核心要素：

① 需要一个 state 变量，标记该锁的状态。state 变量至少有两个值：0、1。对 state 变量的操作，要确保线程安全，也就是会用到 CAS。

② 需要记录当前是哪个线程持有锁。

③ 需要底层支持对一个线程进行阻塞或唤醒操作。

④ 需要有一个队列维护所有阻塞的线程。这个队列也必须是线程安全的无锁队列，也需要用到 CAS。

针对要素①②，在上面两个类中有对应的体现：

```
public abstract class AbstractOwnableSynchronizer{
```

```
    ...
    private transient Thread exclusiveOwnerThread;    //记录锁被哪个线程持有
}
public abstract class AbstractQueuedSynchronizer
    extends AbstractOwnableSynchronizer{
    private volatile int state;           //记录锁的状态,通过 CAS 修改 state 值
    ...
}
```

state 取值不仅可以是 0、1,还可以大于 1,就是为了支持锁的可重入性。例如,同样一个线程,调用 5 次 lock,state 会变成 5;然后调用 5 次 unlock,state 减为 0。

当 state = 0 时,没有线程持有锁,exclusiveOwnerThread = null;

当 state = 1 时,有一个线程持有锁,exclusiveOwnerThread =该线程;

当 state > 1 时,说明该线程重入了该锁。

针对要素③,在 Unsafe 类中,提供了阻塞或唤醒线程的一对操作原语,也就是 park/unpark。

```
    public native void unpark(Object var1);
    public native void park(boolean var1, long var2);
```

有一个 LockSupport 的工具类,对这一对原语做了简单封装:

```
public class LockSupport {
    ...
    public static void park() {
    UNSAFE.park(false, 0);
    }
    public static void unpark(Thread thread) {
    if (thread != null)
        UNSAFE.unpark(thread);
}
}
```

在当前线程中调用 park(),该线程就会被阻塞;在另外一个线程中,调用 unpark(Thread t),传入一个被阻塞的线程,就可以唤醒阻塞在 park()地方的线程。

尤其是 unpark(Thread t),它实现了一个线程对另外一个线程的"精准唤醒"。前面讲到的 wait()/notify(),notify 也只是唤醒某一个线程,但无法指定具体唤醒哪个线程。

针对要素④,在 AQS 中利用双向链表和 CAS 实现了一个阻塞队列。如下所示。

```
public abstract class AbstractQueuedSynchronizer{
    ...
    static final class Node {
```

```
volatile Thread thread;    //每个 Node 对应一个被阻塞的线程
volatile Node prev;
volatile Node next;
...
}
private transient volatile Node head;
private transient volatile Node tail;
...
}
```

阻塞队列是整个 AQS 核心中的核心，下面做进一步的阐述。如图 3-2 所示，head 指向双向链表头部，tail 指向双向链表尾部。入队就是把新的 Node 加到 tail 后面，然后对 tail 进行 CAS 操作；出队就是对 head 进行 CAS 操作，把 head 向后移一个位置。

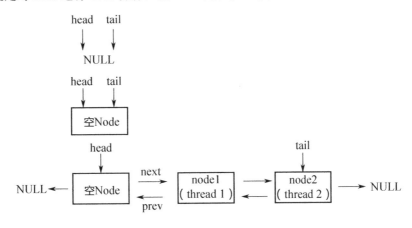

图 3-2　阻塞队列的示意图

初始的时候，head = tail = NULL；然后，在往队列中加入阻塞的线程时，会新建一个空的 Node，让 head 和 tail 都指向这个空 Node；之后，在后面加入被阻塞的线程对象。所以，当 head = tail 的时候，说明队列为空。

3.1.5　公平与非公平的 lock() 的实现差异

下面分析基于 AQS，ReentrantLock 在公平性和非公平性上的实现差异。

```
final static class NonfairSync extends Sync {
...
    final void lock() {
        if(compareAndSetState(0, 1))   //一上来就尝试修改 state 值，也就是抢锁，
                                       //不考虑队列中有没有其他线程在排队，
                                       //是非公平的
```

```
            setExclusiveOwnerThread(Thread.currentThread());
        else
            acquire(1);
    }
    ...
}
final static class FairSync extends Sync {
    ...
    final void lock() {
        acquire(1);         //没有一上来就抢锁，在这个函数内部排队，是公平的
    }
...}
```

acquire()是 AQS 的一个模板方法，如下所示。

```
public final void acquire(int arg) {
    if (!tryAcquire(arg) &&
        acquireQueued(addWaiter(Node.EXCLUSIVE), arg))
        selfInterrupt();
}
```

tryAcquire(..)是一个虚函数，也就是再次尝试拿锁，被 NonfairSync 与 FairSync 分别实现。acquireQueued(..)函数的目的是把线程放入阻塞队列，然后阻塞该线程。

下面再分别来看 FairSync 与 NonfairSync 的 tryAcquire(1)有什么区别。

```
abstract static class Sync {
    final boolean nonfairTryAcquire(int acquires) {
        final Thread current = Thread.currentThread();
        int c = getState();
        if (c == 0) {    //无人持有锁，就开始下面的抢锁
            if (compareAndSetState(0, acquires))  {  //拿锁成功，把
                                                     //excluseThread 设置成
                                                     //当前线程
                setExclusiveOwnerThread(current);
                return true;
            }
        }
        else if (current == getExclusiveOwnerThread()) {  //已经拿到锁，再
                                                          //次重入，直接累
                                                          //加 state 变量
            int nextc = c + acquires;
            if (nextc < 0) // overflow
                throw new Error("Maximum lock count exceeded");
            setState(nextc);
```

```
            return true;
        }
        return false;
    }
    ...
}
static final class FairSync extends Sync {
    protected final boolean tryAcquire(int acquires) {
        final Thread current = Thread.currentThread();
        int c = getState();
        if (c == 0) {
            if (!hasQueuedPredecessors() &&
                compareAndSetState(0, acquires)) {
                setExclusiveOwnerThread(current);
                return true;
            }
        }
        else if (current == getExclusiveOwnerThread()) {
            int nextc = c + acquires;
            if (nextc < 0)
                throw new Error("Maximum lock count exceeded");
            setState(nextc);
            return true;
        }
        return false;
    }
...}
```

这两段代码非常相似,唯一的区别是第二段代码多了一个 if(!hasQueuedPredecessors())。什么意思呢?就是只有当 c==0(没有线程持有锁),并且排在队列的第 1 个时(即当队列中没有其他线程的时候),才去抢锁,否则继续排队,这才叫"公平"!

3.1.6 阻塞队列与唤醒机制

下面进入锁的最为关键的部分,即 acquireQueued(..) 函数内部一探究竟。

```
public final void acquire(int arg) {
    if (!tryAcquire(arg) &&
        acquireQueued(addWaiter(Node.EXCLUSIVE), arg))
        selfInterrupt();
}
```

先说 addWaiter(..) 函数,就是为当前线程生成一个 Node,然后把 Node 放入双向链表的尾部。要注意的是，这只是把 Thread 对象放入了一个队列中而已，线程本身并未阻塞。

```java
private Node addWaiter(Node mode) {
    Node node = new Node(Thread.currentThread(), mode);
    Node pred = tail;
    if (pred != null) {
        node.prev = pred;
        if (compareAndSetTail(pred, node)) {   //先尝试加到队列尾部，如果不成
                                               //功，则执行下面的 enq(node)
            pred.next = node;
            return node;
        }
    }
    enq(node);         //enq 内部会进行队列的初始化，新建一个空的 Node。然后不断尝试自
                       //旋，直至成功地把该 Node 加入队列尾部为止
    return node;
}
```

在 addWaiter(..)函数把 Thread 对象加入阻塞队列之后的工作就要靠 acquireQueued(..)函数完成。线程一旦进入 acquireQueued(..)就会被无限期阻塞，即使有其他线程调用 interrupt()函数也不能将其唤醒，除非其他线程释放了锁，并且该线程拿到了锁，才会从 accquireQueued(..)返回。

注意：进入 acquireQueued(..)，该线程被阻塞。在该函数返回的一刻，就是拿到锁的那一刻，也就是被唤醒的那一刻，此时会删除队列的第一个元素（head 指针前移 1 个节点）。

```java
final boolean acquireQueued(final Node node, int arg) {
    boolean failed = true;
    try {
        boolean interrupted = false;
        for (;;) {
            final Node p = node.predecessor();
            if (p == head && tryAcquire(arg)) {   //被唤醒，如果自己在队列头
                                                  //部(自己的前一个节点是 head
                                                  //指向的空节点)，则尝试拿锁
                setHead(node);   //拿锁成功，出队列（即 head 指针前移一个节点），同
                                 //时会把 node 的 thread 变量置为 NULL，所以 head
                                 //还是指向了一个空节点
                p.next = null;
                failed = false;
                return interrupted;
            }
            if (shouldParkAfterFailedAcquire(p, node) &&
                parkAndCheckInterrupt())      //自己调用 park()阻塞自己
```

```
                interrupted = true;
        }
    } finally {
        if (failed)
            cancelAcquire(node);
    }
}
```

首先，acquireQueued(..)函数有一个返回值，表示什么意思呢？虽然该函数不会中断响应，但它会记录被阻塞期间有没有其他线程向它发送过中断信号。如果有，则该函数会返回 true；否则，返回 false。

基于这个返回值，才有了下面的代码：

```
public final void acquire(int arg) {
    if (!tryAcquire(arg) &&
        acquireQueued(addWaiter(Node.EXCLUSIVE), arg))
        selfInterrupt();
}
static void selfInterrupt() {
    Thread.currentThread().interrupt();
}
```

当 acquireQueued(..) 返回 true 时，会调用 selfInterrupt()，自己给自己发送中断信号，也就是自己把自己的中断标志位设为 true。之所以要这么做，是因为自己在阻塞期间，收到其他线程中断信号没有及时响应，现在要进行补偿。这样一来，如果该线程在 lock 代码块内部有调用 sleep() 之类的阻塞方法，就可以抛出异常，响应该中断信号。

阻塞就发生在下面这个函数中：

```
private final boolean parkAndCheckInterrupt() {
    LockSupport.park(this);
    return Thread.interrupted();
}
```

线程调用 park() 函数，自己把自己阻塞起来，直到被其他线程唤醒，该函数返回。park() 函数返回有两种情况。

情况 1：其他线程调用了 unpark(Thread t)。

情况 2：其他线程调用了 t.interrupt()。这里要注意的是，lock()不能响应中断，但 LockSupport.park() 会响应中断。

也正因为 LockSupport.park() 可能被中断唤醒，acquireQueued(..)函数才写了一个 for 死循环。唤醒之后，如果发现自己排在队列头部，就去拿锁；如果拿不到锁，则再次自己阻塞自己。不断重复此过程,直到拿到锁。

被唤醒之后,通过 Thread.interrupted() 来判断是否被中断唤醒。如果是情况 1,会返回 false;如果是情况 2,则返回 true。

3.1.7 unlock() 实现分析

说完了 lock,下面分析 unlock 的实现。unlock 不区分公平还是非公平。

```
    public void unlock() {
        sync.release(1);
    }
    public final boolean release(int arg) {
        if (tryRelease(arg)) {
            Node h = head;
            if (h != null && h.waitStatus != 0)
                unparkSuccessor(h);
            return true;
        }
        return false;
    }
protected final boolean tryRelease(int releases) {
            int c = getState() - releases;
            if (Thread.currentThread() != getExclusiveOwnerThread())
            //很显然,只有锁的拥有者才有资格调用 unlock() 函数,否则直接抛出异常
                throw new IllegalMonitorStateException();
            boolean free = false;
            if (c == 0) {     //每调用 1 次 tryRelease,state 值减 1,直至减到 0,
                              //才代表锁可以被成功释放
                free = true;
                setExclusiveOwnerThread(null);
            }
            setState(c);   //关键点:没有使用 CAS,而是直接用 set。因为是排他锁,只有 1
                           //个线程能调减 state 值
            return free;
        }

private void unparkSuccessor(Node node) {
…
        Node s = node.next;
        if (s == null || s.waitStatus > 0) {
            s = null;
```

```
            for (Node t = tail; t != null && t != node; t = t.prev)
                if (t.waitStatus <= 0)
                    s = t;
        }
        if (s != null)
            LockSupport.unpark(s.thread);  //关键的一句
    }
```

release()里面做了两件事：tryRelease(..)函数释放锁；unparkSuccessor(..)函数唤醒队列中的后继者。

在上面的代码中有一个关键点要说明：因为是排他锁，只有已经持有锁的线程才有资格调用 release(..)，这意味着没有其他线程与它争抢。所以，在上面的 tryRelease(..)函数中，对 state 值的修改，不需要 CAS 操作，直接减 1 即可。

但对于读写锁中的读锁，也就是 releaseShared(..)，就不一样了，见后续分析。

3.1.8　lockInterruptibly()实现分析

上面的 lock 不能被中断，这里的 lockInterruptibly()可以被中断，下面看一下两者在实现上有什么差别。

```
public void lockInterruptibly() throws InterruptedException {
    sync.acquireInterruptibly(1);
}
public final void acquireInterruptibly(int arg)
        throws InterruptedException {
    if (Thread.interrupted())  throw new InterruptedException();
    if (!tryAcquire(arg))
        doAcquireInterruptibly(arg);
}
```

这里的 acquireInterruptibly(..)也是 AQS 的模板方法，里面的 tryAcquire(..)分别被 FairSync 和 NonfairSync 实现，此处不再重复叙述。这里主要讲 doAcquireInterruptibly(..)函数。

```
private void doAcquireInterruptibly(int arg)
        throws InterruptedException {
    final Node node = addWaiter(Node.EXCLUSIVE);
    boolean failed = true;
    try {
        for (;;) {
            final Node p = node.predecessor();
            if (p == head && tryAcquire(arg)) {
                setHead(node);
```

```
                p.next = null; // help GC
                failed = false;
                return;
            }
            if (shouldParkAfterFailedAcquire(p, node) &&
                parkAndCheckInterrupt())
                throw new InterruptedException();    //关键的一行,收到中断信号,
                                                     //不再阻塞,直接抛异常再返回
        }
    } finally {
        if (failed)
            cancelAcquire(node);
    }
}
```

明白了 accquireQueued(..) 原理,此处就很简单了。当 parkAndCheckInterrupt() 返回 true 的时候,说明有其他线程发送中断信号,直接抛出 InterruptedException,跳出 for 循环,整个函数返回。

3.1.9　tryLock()实现分析

```
public boolean tryLock() {
    return sync.nonfairTryAcquire(1);
}
```

tryLock() 实现基于调用非公平锁的 tryAcquire(..),对 state 进行 CAS 操作,如果操作成功就拿到锁;如果操作不成功则直接返回 false,也不阻塞。

3.2　读写锁

和互斥锁相比,读写锁(ReentrantReadWriteLock)就是读线程和读线程之间可以不用互斥了。在正式介绍原理之前,先看一下相关类的继承体系。

3.2.1　类继承层次

如图 3-3 所示,ReadWriteLock 是一个接口,内部由两个 Lock 接口组成。

```
public interface ReadWriteLock {
    Lock readLock();
    Lock writeLock();
}
```

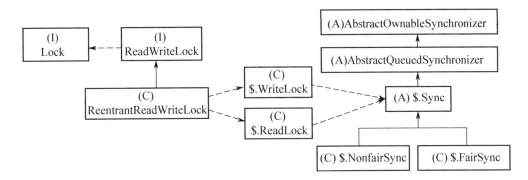

图 3-3 ReentrantReadWriteLock 类继承层次

ReentrantReadWriteLock 实现了该接口，使用方式如下：

```
ReadWriteLock rwLock = new ReentrantReadWriteLock();
Lock rLock = rwLock.readLock();
rLock.lock();
rLock.unlock();
Lock wLock = rwLock.writeLock();
wLock.lock();
wLock.unlock();
```

也就是说，当使用 ReadWriteLock 的时候，并不是直接使用，而是获得其内部的读锁和写锁，然后分别调用 lock/unlock。

3.2.2 读写锁实现的基本原理

从表面来看，ReadLock 和 WriteLock 是两把锁，实际上它只是同一把锁的两个视图而已。什么叫两个视图呢？可以理解为是一把锁，线程分成两类：读线程和写线程。读线程和读线程之间不互斥（可以同时拿到这把锁），读线程和写线程互斥，写线程和写线程也互斥。

从下面的构造函数也可以看出，readerLock 和 writerLock 实际共用同一个 sync 对象。sync 对象同互斥锁一样，分为非公平和公平两种策略，并继承自 AQS。

```
public ReentrantReadWriteLock() {
    this(false);
}
public ReentrantReadWriteLock(boolean fair) {
    sync = fair ? new FairSync() : new NonfairSync();
    readerLock = new ReadLock(this);
    writerLock = new WriteLock(this);
}
```

同互斥锁一样，读写锁也是用 state 变量来表示锁状态的。只是 state 变量在这里的含义和互斥锁完全不同。在内部类 Sync 中，对 state 变量进行了重新定义，如下所示。

```
static abstract class Sync extends AbstractQueuedSynchronizer {
 ...
 static final int SHARED_SHIFT   = 16;
 static final int SHARED_UNIT    = (1 << SHARED_SHIFT);
 static final int MAX_COUNT      = (1 << SHARED_SHIFT) - 1;
 static final int EXCLUSIVE_MASK = (1 << SHARED_SHIFT) - 1;
 //持有读锁的线程的重入次数
 static int sharedCount(int c)    { return c >>> SHARED_SHIFT; }
 //持有写锁的线程的重入次数
 static int exclusiveCount(int c) { return c & EXCLUSIVE_MASK; }
 ...
}
```

也就是把 state 变量拆成两半，低 16 位，用来记录写锁。但同一时间既然只能有一个线程写，为什么还需要 16 位呢？这是因为一个写线程可能多次重入。例如，低 16 位的值等于 5，表示一个写线程重入了 5 次。

高 16 位，用来"读"锁。例如，高 16 位的值等于 5，可以表示 5 个读线程都拿到了该锁；也可以表示一个读线程重入了 5 次。

这个地方的设计很巧妙，为什么要把一个 int 类型变量拆成两半，而不是用两个 int 型变量分别表示读锁和写锁的状态呢？这是因为无法用一次 CAS 同时操作两个 int 变量，所以用了一个 int 型的高 16 位和低 16 位分别表示读锁和写锁的状态。

当 state = 0 时，说明既没有线程持有读锁，也没有线程持有写锁；当 state != 0 时，要么有线程持有读锁，要么有线程持有写锁，两者不能同时成立，因为读和写互斥。这时再进一步通过 sharedCount(state) 和 exclusiveCount(state) 判断到底是读线程还是写线程持有了该锁。

3.2.3 AQS 的两对模板方法

下面介绍在 ReentrantReadWriteLock 的两个内部类 ReadLock 和 WriteLock 中，是如何使用 state 变量的。

```
public static class ReadLock implements Lock, java.io.Serializable {
 ...
 public void lock() {
   sync.acquireShared(1);
 }
 public void unlock(){
   sync.releaseShared(1);
```

```
        }
        ...
    }
    public static class WriteLock implements Lock, java.io.Serializable {
        ...
        public void lock() {
            sync.acquire(1);
        }
        public void unlock() {
            sync.release(1);
        }
        ...
    }
```

　　acquire/release、acquireShared/releaseShared 是 AQS 里面的两对模板方法。互斥锁和读写锁的写锁都是基于 acquire/release 模板方法来实现的，读写锁的读锁是基于 acquireShared/releaseShared 这对模板方法来实现的。这两对模板方法的代码如下：

```
public abstract class AbstractQueuedSynchronizer extends AbstractOwnable-
Synchronizer
    implements java.io.Serializable {
    ...
    public final void acquire(int arg) {
        if (!tryAcquire(arg) &&          //tryAcquire 被各种 Sync 子类实现
            acquireQueued(addWaiter(Node.EXCLUSIVE), arg))
            selfInterrupt();
    }
    public final void acquireShared(int arg) {
        if (tryAcquireShared(arg) < 0)   //tryAcquireShared 被各种 Sync 子类实现
            doAcquireShared(arg);
    }
    public final boolean release(int arg) {
        if (tryRelease(arg)) {           //tryRelease 被各种 Sync 子类实现
            Node h = head;
            if (h != null && h.waitStatus != 0)
                unparkSuccessor(h);
            return true;
        }
        return false;
    }
    public final boolean releaseShared(int arg) {
        if (tryReleaseShared(arg)) {     //tryRelease 被各种 Sync 子类实现
            doReleaseShared();
```

```
            return true;
        }
        return false;
    }
    ...
}
```

将读/写、公平/非公平进行排列组合，就有 4 种组合。如图 3-4 所示，上面的两个函数都是在 Sync 中实现的。Sync 中的两个函数又是模板方法，在 NonfairSync 和 FairSync 中分别有实现。最终的对应关系如下：

（1）读锁的公平实现：Sync.tryAccquireShared() + FairSync 中的两个覆写的子函数。

（2）读锁的非公平实现：Sync.tryAccquireShared() + NonfairSync 中的两个覆写的子函数。

（3）写锁的公平实现：Sync.tryAccquire() + FairSync 中的两个覆写的子函数。

（4）写锁的非公平实现：Sync.tryAccquire() + NonfairSync 中的两个覆写的子函数。

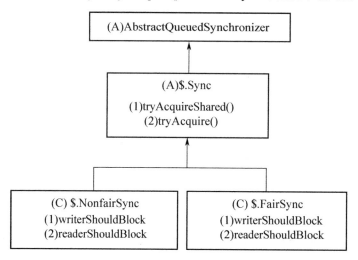

图 3-4　四种锁的策略的实现示意图

```
static final class NonfairSync extends Sync {
    final boolean writerShouldBlock() { //写线程抢锁的时候是否应该阻塞
        return false;                   //写线程在抢锁之前永远不被阻塞，是非公平的
    }
    final boolean readerShouldBlock() { //读线程抢锁的时候是否应该阻塞
        return apparentlyFirstQueuedIsExclusive();  //读线程抢锁的时候，当队列中第 1
                                                    //个元素是写线程的时候，要阻塞
    }
```

```
}
static final class FairSync extends Sync {
    final boolean writerShouldBlock() { //写线程抢锁的时候是否应该阻塞
        return hasQueuedPredecessors(); //写线程在抢锁之前,如果队列里有其他线程在排
                                        //队,就要阻塞,所以是公平的
    }
    final boolean readerShouldBlock() { //读线程抢锁的时候是否应该阻塞
        return hasQueuedPredecessors(); //读线程在抢锁之前,如果队列里有其他线程在排
                                        //队,就要阻塞,所以是公平的
    }
}
```

上面的代码介绍了 ReentrantReadWriteLock 里面的 NonfairSync 和 FairSync 的实现过程,对应了上面的四种实现策略,下面分别解释。

对于公平,比较容易理解,不论是读锁,还是写锁,只要队列中有其他线程在排队(排队等读锁,或者排队等写锁),就不能直接去抢锁,要排在队列尾部。

对于非公平,读锁和写锁的实现策略略有差异。先说写锁,写线程能抢锁,前提是 state = 0,只有在没有其他线程持有读锁或写锁的情况下,它才有机会去抢锁。或者 state != 0,但那个持有写锁的线程是它自己,再次重入。写线程是非公平的,就是不管三七二十一就去抢,即一直返回 false。

但对于读线程,能否也不管三七二十一,上来就去抢呢?不行!因为读线程和读线程是不互斥的,假设当前线程被读线程持有,然后其他读线程还非公平地一直去抢,可能导致写线程永远拿不到锁,所以对于读线程的非公平,要做一些"约束"。当发现队列的第 1 个元素是写线程的时候,读线程也要阻塞一下,不能"肆无忌惮"地直接去抢。

明白策略后,下面具体介绍四种实现方面的差异。

3.2.4 WriteLock 公平与非公平实现

写锁是排他锁,实现策略类似于互斥锁,重写了 tryAcquire/tryRelease 方法。

1. tryAcquire()实现分析

```
protected final boolean tryAcquire(int acquires) {
    Thread current = Thread.currentThread();
    int c = getState();
    int w = exclusiveCount(c);    //写线程只能有一个,但写线程可以多次重入
    if (c != 0) {                 //c != 0 说明有读线程或者写线程持有锁
                                  //(Note: if c != 0 and w == 0 then shared count != 0)
```

```
        if (w == 0 || current != getExclusiveOwnerThread())
        //w == 0，说明锁被读线程持有，只能返回；w != 0，持有写锁的线程不是自己，也只能返回
            return false;
        if (w + exclusiveCount(acquires) > MAX_COUNT)    //16 位用满了，超过了最大重
                                                         //入次数
            throw new Error("Maximum lock count exceeded");
        // Reentrant acquire
        setState(c + acquires);
        return true;
    }
    if (writerShouldBlock() ||     //writerShouldBlock 就是四种不同实现策略
        !compareAndSetState(c, c + acquires))
        return false;
    setExclusiveOwnerThread(current);
    return true;
}
```

把上面的代码拆开进行分析，如下：

（1）if (c != 0) and w == 0，说明当前一定是读线程拿着锁，写锁一定拿不到，返回 false。

（2）if (c != 0) and w != 0，说明当前一定是写线程拿着锁，执行 current != getExclusive-OwnerThread() 的判断，发现 ownerThread 不是自己，返回 false。

（3）c != 0，w != 0，且 current = getExclusiveOwnerThread()，才会走到 if (w + exclusive-Count(acquires) > MAX_COUNT)。判断重入次数，重入次数超过最大值，抛出异常。

因为是用 state 的低 16 位保存写锁重入次数的，所以 MAX_COUNT 是 2^{16}。如果超出这个值，会写到读锁的高 16 位上。为了避免这种情形，这里做了一个检测。当然，一般不可能重入这么多次。

（4）if(c = 0)，说明当前既没有读线程，也没有写线程持有该锁。可以通过 CAS 操作开抢了。

```
if (writerShouldBlock() ||
    !compareAndSetState(c, c + acquires))
```

抢成功后，调用 setExclusiveOwnerThread(current)，把 ownerThread 设成自己。

公平实现和非公平实现几乎一模一样，只是 writerShouldBlock() 分别被 FairSync 和 NonfairSync 实现，在上一节已讲。

2. tryRelease(..)实现分析

```
protected final boolean tryRelease(int releases) {
    if (!isHeldExclusively())
        throw new IllegalMonitorStateException();
```

```
    int nextc = getState() - releases;
    boolean free = exclusiveCount(nextc) == 0;
    if (free)
        setExclusiveOwnerThread(null);
    setState(nextc);   //因为写锁是排他的，在当前线程持有写锁的时候，其他线程既不会持有写锁，也
                       //不会持有读锁。所以，这里对 state 值的调减不需要 CAS 操作，直接减 1 即可
    return free;
}
```

3.2.5 ReadLock 公平与非公平实现

读锁是共享锁，重写了 tryAcquireShared/tryReleaseShared 方法，其实现策略和排他锁有很大的差异。

1. tryAcquireShared(..)实现分析

```
protected final int tryAcquireShared(int unused) {
    Thread current = Thread.currentThread();
    int c = getState();
    if (exclusiveCount(c) != 0 &&        //写锁被某线程持有，并且这个线程还不是自己，读锁
                                         //肯定拿不到，直接返回
        getExclusiveOwnerThread() != current)
        return -1;
    int r = sharedCount(c);
    if (!readerShouldBlock() &&          //公平和非公平的差异就在于这个函数
        r < MAX_COUNT &&
        compareAndSetState(c, c + SHARED_UNIT)) {    //CAS 拿读锁，高 16 位加 1
        if (r == 0) {      //r 之前等于 0，说明这是第一个拿到读锁的线程
            firstReader = current;
            firstReaderHoldCount = 1;
        } else if (firstReader == current) {         //不是第一个
            firstReaderHoldCount++;
        } else {
            HoldCounter rh = cachedHoldCounter;
            if (rh == null || rh.tid != current.getId())
                cachedHoldCounter = rh = readHolds.get();
            else if (rh.count == 0)
                readHolds.set(rh);
            rh.count++;
        }
        return 1;
    }
```

```
        return fullTryAcquireShared(current);   //上面拿读锁失败,进入这个函数不断自旋拿读锁
}
```

下面是关于此代码的解释:

(1)
```
if (exclusiveCount(c) != 0 &&
    getExclusiveOwnerThread() != current)
    return -1;
```

低 16 位不等于 0,说明有写线程持有锁,并且只有当 ownerThread != 自己时,才返回-1。这里面有一个潜台词:如果 current = ownerThread,则这段代码不会返回。这是因为一个写线程可以再次去拿读锁!也就是说,一个线程在持有了 WriteLock 后,再去调用 ReadLock.lock 也是可以的。

(2)上面的 compareAndSetState(c, c + SHARED_UNIT),其实是把 state 的高 16 位加 1(读锁的状态),但因为是在高 16 位,必须把 1 左移 16 位再加 1。

(3) firstReader、cachedHoldConunter 之类的变量,只是一些统计变量,在 ReentrantReadWriteLock 对外的一些查询函数中会用到,例如,查询持有读锁的线程列表,但对整个读写互斥机制没有影响,此处不再展开解释。

2. tryReleaseShared(..)实现分析

```
protected final boolean tryReleaseShared(int unused) {
    Thread current = Thread.currentThread();
    ...
    for (;;) {
        int c = getState();
        int nextc = c - SHARED_UNIT;
        if (compareAndSetState(c, nextc))
            return nextc == 0;
    }
}
```

因为读锁是共享锁,多个线程会同时持有读锁,所以对读锁的释放不能直接减 1,而是需要通过一个 for 循环 + CAS 操作不断重试。这是 tryReleaseShared 和 tryRelease 的根本差异所在。

3.3 Condition

3.3.1 Condition 与 Lock 的关系

Condition 本身也是一个接口,其功能和 wait/notify 类似,如下所示。

```
public interface Condition {
void await() throws InterruptedException;
void awaitUninterruptibly();
void signal();
void signalAll();
…
}
```

在讲多线程基础的时候，强调 wait()/notify()必须和 synchronized 一起使用，Condition 也是如此，必须和 Lock 一起使用。因此，在 Lock 的接口中，有一个与 Condition 相关的接口：

```
public interface Lock {
    void lock();
    void lockInterruptibly() throws InterruptedException;
    boolean tryLock();
    boolean tryLock(long time, TimeUnit unit) throws InterruptedException;
    void unlock();
    Condition newCondition();   //所有的Condition都是从Lock中构造出来的
}
```

3.3.2 Condition 的使用场景

在讲 Condition 的实现原理之前，先以 ArrayBlockingQueue 的实现为例，介绍 Condition 的使用场景。如下所示为一个用数组实现的阻塞队列，执行 put(..)操作的时候，队列满了，生成者线程被阻塞；执行 take()操作的时候，队列为空，消费者线程被阻塞。

```
public class ArrayBlockingQueue<E> extends AbstractQueue<E>
        implements BlockingQueue<E>, java.io.Serializable {
    …
    /** The queued items */
    final Object[] items;
    /** items index for next take, poll, peek or remove */
    int takeIndex;
    /** items index for next put, offer, or add */
    int putIndex;
    /** Number of elements in the queue */
    int count;
    //其核心是一把锁+两个条件
    final ReentrantLock lock;
    private final Condition notEmpty;
    private final Condition notFull;
    …
}
```

```java
public ArrayBlockingQueue(int capacity, boolean fair) {
    if (capacity <= 0)
        throw new IllegalArgumentException();
    this.items = new Object[capacity];
    lock = new ReentrantLock(fair);    //在构造函数中，一把锁两个 Condition
    notEmpty = lock.newCondition();
    notFull =  lock.newCondition();
}
public void put(E e) throws InterruptedException {
    checkNotNull(e);
    final ReentrantLock lock = this.lock;
    lock.lockInterruptibly();
    try {
        while (count == items.length)
            notFull.await();    //put 的时候，队列满了，阻塞于"非满"条件
        insert(e);
    } finally {
        lock.unlock();
    }
}
private void insert(E x) {
    items[putIndex] = x;
    putIndex = inc(putIndex);
    ++count;
    notEmpty.signal();          //put 进去之后，通知非空条件
}
public E take() throws InterruptedException {
    final ReentrantLock lock = this.lock;
    lock.lockInterruptibly();
    try {
        while (count == 0)
            notEmpty.await(); //take 的时候，队列为空，阻塞在"非空"条件
        return extract();
    } finally {
        lock.unlock();
    }
}
private E extract() {
    final Object[] items = this.items;
    E x = this.<E>cast(items[takeIndex]);
    items[takeIndex] = null;
```

```
    takeIndex = inc(takeIndex);
    --count;
    notFull.signal();       //take 操作完成,通知非满条件
    return x;
}
```

3.3.3 Condition 实现原理

可以发现，Condition 的使用很简洁，避免了 wait/notify 的生成者通知生成者、消费者通知消费者的问题，这是如何做到的呢？下面进入 Condition 内部一探究竟。

因为 Condition 必须和 Lock 一起使用，所以 Condition 的实现也是 Lock 的一部分。下面先分别看一下互斥锁和读写锁中 Condition 的构造。

```
public class ReentrantLock implements Lock, java.io.Serializable {
…
public Condition newCondition() {
return sync.newCondition();
}
}

public class ReentrantReadWriteLock implements ReadWriteLock, java.io.Serializable {
…
public static class ReadLock implements Lock, java.io.Serializable {
…
public Condition newCondition() {
throw new UnsupportedOperationException();   //读锁不支持 Condition
}
}
public static class WriteLock implements Lock, java.io.Serializable {
…
public Condition newCondition() {
return sync.newCondition();
}
}
}
```

首先，读写锁中的 ReadLock 是不支持 Condition 的，读写锁的写锁和互斥锁都支持 Condition。虽然它们各自调用的是自己的内部类 Sync，但内部类 Sync 都继承自 AQS。因此，上面的代码 sync.newCondition 最终都调用了 AQS 中的 newCondition。

```
public abstract class AbstractQueuedSynchronizer extends AbstractOwnableSynchronizer
```

```
implements java.io.Serializable {
public class ConditionObject implements Condition, java.io.Serializable {
    … //Condition 的所有实现,都在 ConditionObject 里面
}
final ConditionObject newCondition() {
    return new ConditionObject();
}
}
```

每一个 Condition 对象上面,都阻塞了多个线程。因此,在 ConditionObject 内部也有一个双向链表组成的队列,如下所示。

```
public class ConditionObject implements Condition, java.io.Serializable {
…
    private transient Node firstWaiter;
    private transient Node lastWaiter;
    …
}
```

下面来看一下在 await()/notify() 函数中,是如何使用这个队列的。

3.3.4 await()实现分析

```
public final void await() throws InterruptedException {
    if (Thread.interrupted())                        //正要执行await()操作,收到中断信号,抛
                                                     //出异常
        throw new InterruptedException();
    Node node = addConditionWaiter();                //加入Condition的等待队列
    int savedState = fullyRelease(node);             //关键的一句:阻塞在Condition之前必须
                                                     //先释放锁,否则会死锁
    int interruptMode = 0;
    while (!isOnSyncQueue(node)) {
        LockSupport.park(this);                      //自己阻塞自己
        if ((interruptMode = checkInterruptWhileWaiting(node)) != 0)
            break;
    }
    if (acquireQueued(node, savedState) && interruptMode != THROW_IE)   //重新拿锁
        interruptMode = REINTERRUPT;
    if (node.nextWaiter != null) // clean up if cancelled
        unlinkCancelledWaiters();
    if (interruptMode != 0)
        reportInterruptAfterWait(interruptMode);     //被中断唤醒,向外抛出中断异常
}
```

关于 await,有几个关键点要说明:

（1）线程调用 await()的时候，肯定已经先拿到了锁。所以，在 addConditionWaiter()内部，对这个双向链表的操作不需要执行 CAS 操作，线程天生是安全的，代码如下。

```
private Node addConditionWaiter() {
Node t = lastWaiter;
...
Node node = new Node(Thread.currentThread(), Node.CONDITION);
if (t == null)
    firstWaiter = node;
else
    t.nextWaiter = node;
lastWaiter = node;
return node;
}
```

（2）在线程执行 wait 操作之前，必须先释放锁。也就是 fullyRelease(node)，否则会发生死锁。这个和 wait/notify 与 synchronized 的配合机制一样。

（3）线程从 wait 中被唤醒后，必须用 acquireQueued(node, savedState)函数重新拿锁。

（4）checkInterruptWhileWaiting(node)代码在 park(this)代码之后，是为了检测在 park 期间是否收到过中断信号。当线程从 park 中醒来时，有两种可能：一种是其他线程调用了 unpark，另一种是收到中断信号。这里的 await()函数是可以响应中断的，所以当发现自己是被中断唤醒的，而不是被 unpark 唤醒的时，会直接退出 while 循环，await()函数也会返回。

（5）isOnSyncQueue(node)用于判断该 Node 是否在 AQS 的同步队列里面。初始的时候，Node 只在 Condition 的队列里，而不在 AQS 的队列里。但执行 notity 操作的时候，会放进 AQS 的同步队列。

3.3.5　awaitUninterruptibly()实现分析

与 await()不同，awaitUninterruptibly()不会响应中断，其函数的定义中不会有中断异常抛出，下面分析其实现和 await()的区别。

```
public final void awaitUninterruptibly() {
Node node = addConditionWaiter();
int savedState = fullyRelease(node);
boolean interrupted = false;
while (!isOnSyncQueue(node)) {
    LockSupport.park(this);
    if (Thread.interrupted())     //从 park 中醒来，收到中断，不退出，继续执行 while 循环
```

```
            interrupted = true;
    }
    if (acquireQueued(node, savedState) || interrupted)
        selfInterrupt();
}
```

可以看出，整体代码和 await()类似，区别在于收到异常后，不会抛出异常，而是继续执行 while 循环。

3.3.6 notify()实现分析

```
public final void signal() {
    if (!isHeldExclusively())                    //只有持有锁的线程，才有资格调用 signal()
        throw new IllegalMonitorStateException();
    Node first = firstWaiter;
    if (first != null)
        doSignal(first);
}
private void doSignal(Node first) {     //唤醒队列中的第 1 个线程
    do {
        if ( (firstWaiter = first.nextWaiter) == null)
            lastWaiter = null;
        first.nextWaiter = null;
    } while (!transferForSignal(first) &&
         (first = firstWaiter) != null);
}
final boolean transferForSignal(Node node) {
    if (!compareAndSetWaitStatus(node, Node.CONDITION, 0))
        return false;
    Node p = enq(node); //这里关键的一句：先把 Node 放入互斥锁的同步队列里，再调用下面的 unpark
    int ws = p.waitStatus;
    if (ws > 0 || !compareAndSetWaitStatus(p, ws, Node.SIGNAL))
        LockSupport.unpark(node.thread);
    return true;
}
```

同 await()一样，在调用 notify()的时候，必须先拿到锁（否则就会抛出上面的异常），是因为前面执行 await()的时候，把锁释放了。

然后，从队列中取出 firstWait，唤醒它。在通过调用 unpark 唤醒它之前，先用 enq(node)函数把这个 Node 放入 AQS 的锁对应的阻塞队列中。也正因为如此，才有了 await()函数里面的判断条件 while (!isOnSyncQueue(node))，这个判断条件被满足，说明 await 线程不是被中断，而是被 unpark 唤醒的。

知道了 notify() 实现原理，notifyAll() 与此类似，此处不再赘述。

3.4 StampedLock

3.4.1 为什么引入 StampedLock

在 JDK 8 中新增了 StampedLock，有了读写锁，为什么还要引入 StampedLock 呢？来看一下表 3-1 的对比。

表 3-1 三种锁的并发度的对比

锁	并发度
ReentrantLock	读与读互斥，读与写互斥，写与写互斥
ReentrantReadWriteLock	读与读不互斥，读与写互斥，写与写互斥
StampedLock	读与读不互斥，读与写不互斥，写与写互斥

可以看到，从 ReentrantLock 到 StampedLock，并发度依次提高。StampedLock 是如何做到"读"与"写"也不互斥、并发地访问的呢？在《软件架构设计：大型网站技术架构与业务架构融合之道》中，谈到 MySQL 高并发的核心机制 MVCC，也就是一份数据多个版本，此处的 StampedLock 有异曲同工之妙。

另一方面，因为 ReentrantLock 采用的是"悲观读"的策略，当第一个读线程拿到锁之后，第二个、第三个读线程还可以拿到锁，使得写线程一直拿不到锁，可能导致写线程"饿死"。虽然在其公平或非公平的实现中，都尽量避免这种情形，但还有可能发生。StampedLock 引入了"乐观读"策略，读的时候不加读锁，读出来发现数据被修改了，再升级为"悲观读"，相当于降低了"读"的地位，把抢锁的天平往"写"的一方倾斜了一下，避免写线程被饿死。

3.4.2 使用场景

在剖析其原理之前，下面先以官方的一个例子来看一下 StampedLock 如何使用。

```
class Point {
    private double x, y;
    private final StampedLock sl = new StampedLock();
    void move(double deltaX, double deltaY) {    //多个线程调用该函数，修改 x、y
                                                  //的值
        long stamp = sl.writeLock();
        try {
```

```
            x += deltaX;
            y += deltaY;
        } finally {
            sl.unlockWrite(stamp);
        }
    }
    double distanceFromOrigin() {                //多个线程调用该函数,求距离
        long stamp = sl.tryOptimisticRead();     //使用"乐观读"
        double currentX = x, currentY = y;       //将共享变量拷贝到线程栈
        if (!sl.validate(stamp)) {               //读的期间有其他线程修改数据,
                                                 //读的是脏数据,放弃
            stamp = sl.readLock();               //升级为"悲观读"
            try {
                currentX = x;
                currentY = y;
            } finally {
                sl.unlockRead(stamp);
            }
        }
        return Math.sqrt(currentX * currentX + currentY * currentY);
    }
...
}
```

如上面代码所示,有一个 Point 类,多个线程调用 move()函数,修改坐标;还有多个线程调用 distanceFromOrigin()函数,求距离。

首先,执行 move 操作的时候,要加写锁。这个用法和 ReadWriteLock 的用法没有区别,写操作和写操作也是互斥的。关键在于读的时候,用了一个"乐观读" sl.tryOptimisticRead(),相当于在读之前给数据的状态做了一个"快照"。然后,把数据拷贝到内存里面,在用之前,再比对一次版本号。如果版本号变了,则说明在读的期间有其他线程修改了数据。读出来的数据废弃,重新获取读锁。关键代码就是下面这三行:

```
long stamp = sl.tryOptimisticRead();     //在读之前,获取数据的版本号
double currentX = x, currentY = y;       //读:将一份数据拷贝到线程的栈内存中
if (!sl.validate(stamp)) {               //读之后:将读之前的版本号和当前的版本号
                                         //比对,判断读出来数据是否可以使用(所谓
                                         //可以使用,是指读的期间没有其他线程修改
                                         //过数据)
```

要说明的是,这三行关键代码对顺序非常敏感,不能有重排序。因为 state 变量已经是 volatile,所以可以禁止重排序,但 stamp 并不是 volatile 的。为此,在 validate(stamp)函数里面插入内存屏障。

3.4.3 "乐观读"的实现原理

首先，StampedLock 是一个读写锁，因此也会像读写锁那样，把一个 state 变量分成两半，分别表示读锁和写锁的状态。同时，它还需要一个数据的 version。但正如前面所说，一次 CAS 没有办法操作两个变量，所以这个 state 变量本身同时也表示了数据的 version。下面先分析 state 变量。

```
public class StampedLock implements java.io.Serializable {
  private static final int LG_READERS = 7;

private static final long RUNIT = 1L;
private static final long WBIT  = 1L << LG_READERS;    //第 8 位表示写锁
private static final long RBITS = WBIT - 1L;           //最低的 7 位表示读锁
private static final long RFULL = RBITS - 1L;          //读锁的数目
private static final long ABITS = RBITS | WBIT;        //读锁和写锁的状态合到一起
private static final long SBITS = ~RBITS;

// Initial value for lock state; avoid failure value zero
private static final long ORIGIN = WBIT << 1;          //state 的初始值
  private transient volatile long state;
  …}
```

结合代码和图 3-5：用最低的 8 位表示读和写的状态，其中第 8 位表示写锁的状态，最低的 7 位表示读锁的状态。因为写锁只有一个 bit 位，所以写锁是不可重入的。

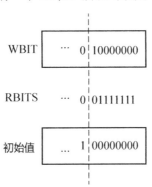

图 3-5 state 变量在不同状态下的取值

初始值不为 0，而是把 WBIT 向左移动了一位，也就是上面的 ORIGIN 常量，构造函数如下所示。

```
public StampedLock() {
  state = ORIGIN;
}
```

为什么 state 的初始值不设为 0 呢？这就要从乐观锁的实现说起。

```
public long tryOptimisticRead() {
long s;
return (((s = state) & WBIT) == 0L) ? (s & SBITS) : 0L;
}
public boolean validate(long stamp) {
U.loadFence();
return (stamp & SBITS) == (state & SBITS);   //当 stamp = 0 时，validate 永远返回 false
}
```

上面两个函数必须结合起来看：当 state&WBIT != 0 的时候，说明有线程持有写锁，上面的 tryOptimisticRead 会永远返回 0。这样，再调用 validate(stamp)，也就是 validate(0) 也会永远返回 false。这正是我们想要的逻辑：当有线程持有写锁的时候，validate 永远返回 false，无论写线程是否释放了写锁。因为无论是否释放了（state 回到初始值）写锁，state 值都不为 0，所以 validate(0) 永远为 false。

为什么上面的 validate(..)函数不直接比较 stamp = state，而要比较 state & SBITS = state & SBITS 呢？因为读锁和读锁是不互斥的！所以，即使在"乐观读"的时候，state 值被修改了，但如果它改的是第 7 位，validate(..)还是会返回 true。

另外要说明的一点是，上面使用了内存屏障 U.loadFence()，是因为在这行代码的下一行里面的 stamp、SBITS 变量不是 volatile 的，由此可以禁止其和前面的 currentX = X，currentY = Y 进行重排序。

通过上面的分析，可以发现 state 的设计非常巧妙。只通过一个变量，既实现了读锁、写锁的状态记录，还实现了数据的版本号的记录。

3.4.4　悲观读/写："阻塞"与"自旋"策略实现差异

同 ReadWriteLock 一样，StampedLock 也要进行悲观的读锁和写锁操作。不过，它不是基于 AQS 实现的，而是内部重新实现了一个阻塞队列。如下所示。

```
public class StampedLock implements java.io.Serializable {
    ...
/** Wait nodes */
static final class WNode {
    volatile WNode prev;
    volatile WNode next;
    volatile WNode cowait;
    volatile Thread thread;
```

```
        volatile int status;            // 取值: 0, WAITING 或 CANCELLED
        final int mode;                 // 取值: RMODE 或 WMODE
        WNode(int m, WNode p) { mode = m; prev = p; }
    }

private transient volatile WNode whead;

private transient volatile WNode wtail;
        ...
    }
```

这个阻塞队列和 AQS 里的很像。刚开始的时候，whead=wtail=NULL，然后初始化，建一个空节点，whead 和 wtail 都指向这个空节点，之后往里面加入一个个读线程或写线程节点。但基于这个阻塞队列实现的锁的调度策略和 AQS 很不一样，也就是"自旋"。在 AQS 里面，当一个线程 CAS state 失败之后，会立即加入阻塞队列，并且进入阻塞状态。但在 StampedLock 中，CAS state 失败之后，会不断自旋，自旋足够多的次数之后，如果还拿不到锁，才进入阻塞状态。为此，根据 CPU 的核数，定义了自旋次数的常量值。如果是单核的 CPU，肯定不能自旋，在多核情况下，才采用自旋策略。

```
    private static final int NCPU = Runtime.getRuntime().availableProcessors();

/** Maximum number of retries before enqueuing on acquisition */
private static final int SPINS = (NCPU > 1) ? 1 << 6 : 0;
```

下面以写锁的加锁，也就是 StampedLock 的 writeLock() 函数为例，来看一下自旋的实现。

```
    public long writeLock() {
    long s, next;
    return ((((s = state) & ABITS) == 0L &&
            U.compareAndSwapLong(this, STATE, s, next = s + WBIT)) ?
            next : acquireWrite(false, 0L));
}
```

如上面代码所示，当 state & ABITS == 0 的时候，说明既没有线程持有读锁，也没有线程持有写锁，此时当前线程才有资格通过 CAS 操作 state。若操作不成功，则调用 acquireWrite() 函数进入阻塞队列，并进行自旋，这个函数是整个加锁操作的核心，代码如下。

```
    private long acquireWrite(boolean interruptible, long deadline) {
    WNode node = null, p;
    for (int spins = -1;;) {  // 入队列时自旋
        long m, s, ns;
        if ((m = (s = state) & ABITS) == 0L) {
            if (U.compareAndSwapLong(this, STATE, s, ns = s + WBIT))
                return ns;     //自旋的时候拿到了锁，函数返回
```

```java
        }
        else if (spins < 0)
            spins = (m == WBIT && wtail == whead) ? SPINS : 0;
        else if (spins > 0) {
            if (LockSupport.nextSecondarySeed() >= 0)
                --spins;          //不断自旋,以一定的概率把spins值往下累减
        }
        else if ((p = wtail) == null) { // 初始化队列
            WNode hd = new WNode(WMODE, null);
            if (U.compareAndSwapObject(this, WHEAD, null, hd))
                wtail = hd;
        }
        else if (node == null)
            node = new WNode(WMODE, p);
        else if (node.prev != p)
            node.prev = p;
        else if (U.compareAndSwapObject(this, WTAIL, p, node)) {
            p.next = node;
            break; //for循环唯一的break,CAS tail 成功(成功加入队列尾部),才会退出for循环
        }
    }

    for (int spins = -1;;) {
        WNode h, np, pp; int ps;
        if ((h = whead) == p) {
            if (spins < 0)
                spins = HEAD_SPINS;
            else if (spins < MAX_HEAD_SPINS)
                spins <<= 1;
            for (int k = spins;;) {
                long s, ns;
                if (((s = state) & ABITS) == 0L) {   //再次尝试拿锁
                    if (U.compareAndSwapLong(this, STATE, s,
                                             ns = s + WBIT)) {
                        whead = node;
                        node.prev = null;
                        return ns;
                    }
                }
                else if (LockSupport.nextSecondarySeed() >= 0 &&
                         --k <= 0)   //不断自旋
                    break;
            }
        }
        else if (h != null) { // help release stale waiters
```

```
            WNode c; Thread w;
            while ((c = h.cowait) != null) { //自己从阻塞中唤醒，然后唤醒cowait中的
                                             //所有reader线程
                if (U.compareAndSwapObject(h, WCOWAIT, c, c.cowait) &&
                    (w = c.thread) != null)
                    U.unpark(w);
            }
        }
        if (whead == h) {
            if ((np = node.prev) != p) {
                if (np != null)
                    (p = np).next = node;
            }
            else if ((ps = p.status) == 0)
                U.compareAndSwapInt(p, WSTATUS, 0, WAITING);
            else if (ps == CANCELLED) {
                if ((pp = p.prev) != null) {
                    node.prev = pp;
                    pp.next = node;
                }
            }
            else {
                long time;
                if (deadline == 0L)
                    time = 0L;
                else if ((time = deadline - System.nanoTime()) <= 0L)
                    return cancelWaiter(node, node, false);
                Thread wt = Thread.currentThread();
                U.putObject(wt, PARKBLOCKER, this);
                node.thread = wt;
                if (p.status < 0 && (p != h || (state & ABITS) != 0L) &&
                    whead == h && node.prev == p)
                    U.park(false, time);    //进入阻塞状态，之后被另外一个线程release唤
                                            //醒，接着往下执行这个for循环
                node.thread = null;
                U.putObject(wt, PARKBLOCKER, null);
                if (interruptible && Thread.interrupted())
                    return cancelWaiter(node, node, true);
            }
        }
    }
}
```

整个acquireWrite(..)函数是两个大的for循环，内部实现了非常复杂的自旋策略。在第一个

大的 for 循环里面，目的就是把该 Node 加入队列的尾部，一边加入，一边通过 CAS 操作尝试获得锁。如果获得了，整个函数就会返回；如果不能获得锁，会一直自旋，直到加入队列尾部。

在第二个大的 for 循环里，也就是该 Node 已经在队列尾部了。这个时候，如果发现自己刚好也在队列头部，说明队列中除了空的 Head 节点，就是当前线程了。此时，再进行新一轮的自旋，直到达到 MAX_HEAD_SPINS 次数，然后进入阻塞。这里有一个关键点要说明：当 release(..) 函数被调用之后，会唤醒队列头部的第 1 个元素，此时会执行第二个大的 for 循环里面的逻辑，也就是接着 for 循环里面 park()函数后面的代码往下执行。

另外一个不同于 AQS 的阻塞队列的地方是，在每个 WNode 里面有一个 cowait 指针，用于串联起所有的读线程。例如，队列尾部阻塞的是一个读线程 1，现在又来了读线程 2、3，那么会通过 cowait 指针，把 1、2、3 串联起来。1 被唤醒之后，2、3 也随之一起被唤醒，因为读和读之间不互斥。

明白加锁的自旋策略后，下面来看锁的释放操作。和读写锁的实现类似，也是做了两件事情：一是把 state 变量置回原位，二是唤醒阻塞队列中的第一个节点。节点被唤醒之后，会继续执行上面的第二个大的 for 循环，自旋拿锁。如果成功拿到，则出队列；如果拿不到，则再次进入阻塞，等待下一次被唤醒。

```
public void unlockWrite(long stamp) {
WNode h;
if (state != stamp || (stamp & WBIT) == 0L)
    throw new IllegalMonitorStateException();
state = (stamp += WBIT) == 0L ? ORIGIN : stamp;   //释放锁，把 state 回归原位
if ((h = whead) != null && h.status != 0)          //唤醒队列头部的第一个节点
    release(h);
}
private void release(WNode h) {
if (h != null) {
   WNode q; Thread w;
   U.compareAndSwapInt(h, WSTATUS, WAITING, 0);
   if ((q = h.next) == null || q.status == CANCELLED) {
       for (WNode t = wtail; t != null && t != h; t = t.prev)
           if (t.status <= 0)
               q = t;
   }
   if (q != null && (w = q.thread) != null)
       U.unpark(w);
  }
}
```

第 4 章 同步工具类

除了锁与 Condition，Concurrent 包还提供了一系列同步工具类。这些同步工具类的原理，有些也是基于 AQS 的，有些则需要特殊的实现机制，这一章将对所有同步工具类的实现原理进行剖析。

4.1 Semaphore

Semaphore 也就是信号量，提供了资源数量的并发访问控制，其使用代码很简单，如下所示。

```
Semaphore available = new Semaphore(10, true);  //初始有10个共享资源。第2个参
                                                //数是公平或非公平选项
available.acquire();       //每次获取一个，如果获取不到，则线程就会阻塞
available.release();       //用完释放
```

如图 4-1 所示，假设有 n 个线程来获取 Semaphore 里面的资源（n > 10），n 个线程中只有 10 个线程能获取到，其他线程都会阻塞。直到有线程释放了资源，其他线程才能获取到。

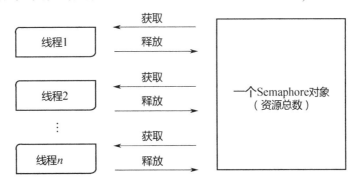

图 4-1 多个线程访问 Semaphore 示意图

当初始的资源个数为 1 的时候，Semaphore 退化为排他锁。正因为如此，Semaphone 的实现原理和锁十分类似，是基于 AQS，有公平和非公平之分。Semaphore 相关类的继承体系如图 4-2 所示。

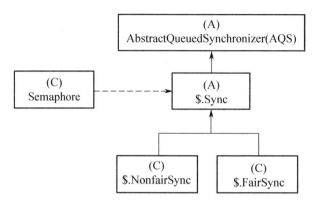

图 4-2　Semaphore 相关类的继承体系

```
public void acquire() throws InterruptedException {
    sync.acquireSharedInterruptibly(1);
}
public void release() {
    sync.releaseShared(1);
}
```

由于 Semaphore 和锁的实现原理基本相同，上面的代码不再展开解释。资源总数即 state 的初始值，在 acquire 里对 state 变量进行 CAS 减操作，减到 0 之后，线程阻塞；在 release 里对 state 变量进行 CAS 加操作。

4.2　CountDownLatch

4.2.1　CountDownLatch 使用场景

考虑一个场景：一个主线程要等待 10 个 Worker 线程工作完毕才退出，就能使用 CountDownLatch 来实现。

```
CountDownLatch doneSignal = new CountDownLatch(10);   //初始为10
doneSignal.await();         //主线程调用该方法，阻塞在这
doneSignal.countDown();  //10个Worker线程 每个线程工作完毕之后 调用1次countDown(),
                         //计算器减1。当减到0之后，主线程就会被唤醒
```

图 4-3 所示为 CountDownLatch 相关类的继承层次，CountDownLatch 原理和 Semaphore 原

理类似，同样是基于 AQS，不过没有公平和非公平之分。

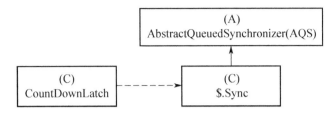

图 4-3　CountDownLatch 相关类的继承层次

4.2.2　await()实现分析

如下所示，await() 调用的是 AQS 的模板方法，这个方法在前面已经介绍过。CountDownLatch.Sync 重新实现了 tryAccuqireShared 方法。

```
public void await() throws InterruptedException {
    this.sync.acquireSharedInterruptibly(1);    //AQS 的模板方法
}
public final void acquireSharedInterruptibly(int arg)
        throws InterruptedException {
    if (Thread.interrupted())
        throw new InterruptedException();
    if (tryAcquireShared(arg) < 0)              //被 CountDownLatch.Sync 重新实现
        doAcquireSharedInterruptibly(arg);
}
protected int tryAcquireShared(int acquires) {
    return (getState() == 0) ? 1 : -1;
}
```

从 tryAcquireShared(..)方法的实现来看，只要 state != 0，调用 await()方法的线程便会被放入 AQS 的阻塞队列，进入阻塞状态。

4.2.3　countDown()实现分析

```
public void countDown() {
    sync.releaseShared(1);
}
//AQS 的模板方法
public final boolean releaseShared(int arg) {
    if (tryReleaseShared(arg)) {    //被 CountDownLatch.Sync 实现
        doReleaseShared();
        return true;
```

```
    }
    return false;
}
protected boolean tryReleaseShared(int releases) {
    for (;;) {
        int c = getState();
        if (c == 0)
            return false;
        int nextc = c-1;
        if (compareAndSetState(c, nextc))
            return nextc == 0;
    }
}
```

countDown() 调用的 AQS 的模板方法 releaseShared()，里面的 tryReleaseShared(..) 被 CountDownLatch.Sync 重新实现。从上面的代码可以看出，只有 state = 0，tryReleaseShared(..) 才会返回 true，然后执行 doReleaseShared(..)，一次性唤醒队列中所有阻塞的线程。

最后做一下小结：因为是基于 AQS 阻塞队列来实现的，所以可以让多个线程都阻塞在 state = 0 条件上，通过 countDown() 一直累减 state，减到 0 后一次性唤醒所有线程。如图 4-4 所示，假设初始总数为 M，N 个线程 await()，M 个线程 countDown()，减到 0 之后，N 个线程被唤醒。

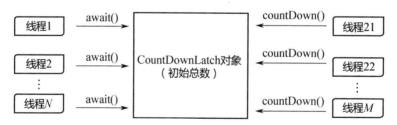

图 4-4　多个线程阻塞在 await() 示意图

4.3　CyclicBarrier

4.3.1　CyclicBarrier 使用场景

CyclicBarrier 使用代码也很简单，如下所示。

```
CyclicBarrier cb = new CyclicBarrier(10);
cb.await();
```

考虑这样一个场景：10 个工程师一起来公司应聘，招聘方式分为笔试和面试。首先，要等人到齐后，开始笔试；笔试结束之后，再一起参加面试。把 10 个人看作 10 个线程，10 个线程之间的同步过程如图 4-5 所示。

图 4-5　10 个线程之间的同步过程

在整个过程中，有 2 个同步点：第 1 个同步点，要等所有应聘者都到达公司，再一起开始笔试；第 2 个同步点，要等所有应聘者都结束笔试，之后一起进入面试环节。具体到每个线程的 run()方法中，就是下面的伪代码：

```
run()
{
report("我已到公司");      //阶段 1
cb.await();               //第 1 个同步点
笔试();                   //阶段 2
cb.await();               //第 2 个同步点
面试();                   //阶段 3
}
```

4.3.2　CyclicBarrier 实现原理

CyclicBarrier 基于 ReentrantLock + Condition 实现。

```
public class CyclicBarrier {
    private final ReentrantLock lock = new ReentrantLock();
    private final Condition trip = lock.newCondition();   //用于线程之间互相唤醒
    private final int parties;    //总线程数
    private int count;            //在讲 await 实现的时候,再详细介绍 count 和 generation
private Generation generation = new Generation();
…
}
```

下面详细介绍 CyclicBarrier 的实现原理。先从构造函数说起，可以看到，不仅可以传入参与方的总数量,还可以传入一个回调函数。当所有的线程被唤醒时,**barrierAction** 被执行。

```java
public CyclicBarrier(int parties, Runnable barrierAction) {
    if (parties <= 0) throw new IllegalArgumentException();
    this.parties = parties;
    this.count = parties;
    this.barrierCommand = barrierAction;
}
```

接下来看一下 await() 函数的实现过程。

```java
public int await() throws InterruptedException, BrokenBarrierException {
    try {
        return dowait(false, 0L);
    } catch (TimeoutException toe) {
        throw new Error(toe);
    }
}

private int dowait(boolean timed, long nanos)
        throws InterruptedException, BrokenBarrierException, TimeoutException
{
    final ReentrantLock lock = this.lock;
    lock.lock();
    try {
            final Generation g = generation;
            if (g.broken)
                throw new BrokenBarrierException();
            if (Thread.interrupted()) {          //响应中断
                breakBarrier();                   //唤醒所有阻塞的线程
                throw new InterruptedException();
            }
            int index = --count;                 //每个线程调用 1 次 await(),count--
            if (index == 0) {                    //当 count 减到 0 的时候,此线程唤醒其他所
                                                 //有线程
                boolean ranAction = false;
                try {
                    final Runnable command = barrierCommand;
                    if (command != null)         //一起唤醒之后,还可以执行一个回调
                        command.run();
                    ranAction = true;
                    nextGeneration();            //唤醒其他所有线程,同时把 count 值复原,
                                                 //用于下一次的 CyclicBarrier(因为
                                                 //CyclicBarrier 可以重复使用)
                    return 0;
```

```
            } finally {
                if (!ranAction)
                    breakBarrier();
            }
        }
        for (;;) {   //当count > 0时, 说明人没到齐, 阻塞自己
            try {
                if (!timed)
                    trip.await();    //关键点: await在阻塞自己的同时, 会把锁释放
                                     //掉, 这样别的线程就会拿到锁, 执行上面的
                                     //index = count--
                else if (nanos > 0L)
                    nanos = trip.awaitNanos(nanos);
            } catch (InterruptedException ie) {
                if (g == generation && ! g.broken) {
                    breakBarrier();
                    throw ie;
                } else {
                    Thread.currentThread().interrupt();
                }
            }
            if (g.broken)
                throw new BrokenBarrierException();
            if (g != generation)    //从阻塞中唤醒, 返回
                return index;
            if (timed && nanos <= 0L) {
                breakBarrier();
                throw new TimeoutException();
            }
        }
    } finally {
        lock.unlock();
    }
}
private void nextGeneration() {
    trip.signalAll();
    count = parties;
    generation = new Generation();
}
private void breakBarrier() {
generation.broken = true;
count = parties;
```

```
    trip.signalAll();
}
```

关于上面的函数,有几点说明:

(1) CyclicBarrier 是可以被重用的。以上一节的应聘场景为例,来了 10 个线程,这 10 个线程互相等待,到齐后一起被唤醒,各自执行接下来的逻辑;然后,这 10 个线程继续互相等待,到齐后再一起被唤醒。每一轮被称为一个 Generation,就是一次同步点。

(2) CyclicBarrier 会响应中断。10 个线程没有到齐,如果有线程收到了中断信号,所有阻塞的线程也会被唤醒,就是上面的 breakBarrier()函数。然后 count 被重置为初始值(parties),重新开始。

(3) 上面的回调函数,barrierAction 只会被第 10 个线程执行 1 次(在唤醒其他 9 个线程之前),而不是 10 个线程每个都执行 1 次。

4.4 Exchanger

4.4.1 使用场景

Exchanger 用于线程之间交换数据,其使用代码很简单,是一个 exchange(..)函数,使用示例如下:

```
Exchange<String> exchange = new Exchange<String>();
//建 1 个多个线程共用的 exchange 对象
//把 exchange 对象传给 4 个线程对象。每个线程在自己的 run 函数里面,调用 exchange,把自己
//的数据当参数传进去,返回值是另外一个线程调用 exchange 传进去的参数
ThreadA a = new ThreadA(exchange);
run(){
   String other = exchange.exchange(self.data)   //如果没有其他线程调用exchange,
                                                 //exchange 自己会阻塞在这。直
                                                 //到有别的线程调用 exchange
}
ThreadB b = new ThreadB(exchange);
run(){
   String other = exchange.exchange(self.data)
}
ThreadC c = new ThreadC(exchange);
run(){
   String other = exchange.exchange(self.data)
}
```

```
ThreadD d = new ThreadD(exchange);
run(){
    String other = exchange.exchange(self.data)
}
```

在上面的例子中，4 个线程并发地调用 exchange(..)，会两两交互数据，如 A/B、C/D、A/C、B/D、A/D 或 B/C。

4.4.2 实现原理

Exchanger 的核心机制和 Lock 一样，也是 CAS + park/unpark。在实现上面，JDK7 和 JDK8 有一定差异，这里以 JDK7 的实现为例进行分析。

首先，在 Exchanger 内部，有两个内部类：Slot 和 Node，代码如下：

```
public class Exchanger<V> {
private static final class Node extends AtomicReference<Object> {
public final Object item;          //自己要拿去交互的数据
public volatile Thread waiter;     //自己

public Node(Object item) {
    this.item = item;
}
}
private static final class Slot extends AtomicReference<Object> {
// 缓存行填充
long q0, q1, q2, q3, q4, q5, q6, q7, q8, q9, qa, qb, qc, qd, qe;
}
...
}
```

每个线程在调用 exchange(..)函数交换数据的时候，会先创建一个 Node 对象，这个 Node 对象就是对该线程的包装，里面包含了 2 个字段：1 个是该线程要交互的数据，另 1 个是该线程自身。这里有个关键点：Node 本身是继承自 AtomicReference 的，所以除了这 2 个字段，Node 还有第 3 个字段，记录的是对方所要交换的数据，初始为 NULL。

Slot 其实是一个 AtomicReference，里面的 q0, q1, ..., qd 变量都是多余的。为什么要添加这些多余的变量呢？是为了优化性能而做的缓存行填充。

Slot 的 AtomicReference 就是指向的一个 Node，通过 Slot 和 Node 相结合，实现了 2 个线程之间的数据交换，如图 4-6 所示。线程 1 持有数据 item1，线程 2 持有数据 item2，各自调用 exchange(..)，会各自生成一个 Node。而 Slot 只会指向 2 个 Node 中的 1 个：如果是线程 1 先调用的 exchange(..)，那么 Slot 就指向 Node1，线程 1 阻塞，等待线程 2 来交换；反之，如果是线程 2 先调用的 exchange(..)，那么 Slot 就指向 Node2，线程 2 阻塞，等待线程 1 来交换数据。

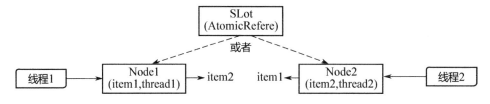

图 4-6　Slot 与 Node 相结合实现 2 个线程交换数据

一个 Slot 只能支持 2 个线程之间交换数据，要实现多个线程并行地交换数据，需要多个 Slot，因此在 Exchanger 里面定义了 Slot 数组：

```
private volatile Slot[] arena = new Slot[CAPACITY];
```

4.4.3　exchange(V x)实现分析

明白了大致思路，下面来看 exchange(V x)函数的详细实现：

```
public V exchange(V x) throws InterruptedException {
    if (!Thread.interrupted()) {
        Object v = doExchange((x == null) ? NULL_ITEM : x, false, 0);
        if (v == NULL_ITEM)
            return null;
        if (v != CANCEL)
            return (V)v;
        Thread.interrupted();
    }
    throw new InterruptedException();
}
```

exchange(..)函数，参数是自己要交互的数据 item，返回的是从对方线程得到的 item。内部调用的是下面的 doExchange(..)函数：

```
private Object doExchange(Object item, boolean timed, long nanos) {
    Node me = new Node(item);          //为该线程新建一个 Node
    int index = hashIndex();           //根据 thread id 计算出自己的交易位置（slot）
    int fails = 0;

    for (;;) {
        Object y;
        Slot slot = arena[index];
        if (slot == null)
```

```
                    createSlot(index);   //slot = null,创建一个slot,然后回到for循环,再次
                                         //开始
            else if ((y = slot.get()) != null &&       //如果slot里面有线程等待(有Node),
                                                       //则尝试与其交换
                    slot.compareAndSet(y, null)) {
                Node you = (Node)y;
                if (you.compareAndSet(null, item)) {   //把Node的AtomiReference指向自
                                                       //己的item,也就是把自己的item给对方
                    LockSupport.unpark(you.waiter);    //唤醒对方
                    return you.item;                   //取回对方的item,返回
                }
            }
            else if (y == null &&                //如果slot里面为空(没有Node),则尝试把自
                                                 //己的Node放进去
                    slot.compareAndSet(null, me)) {
                if (index == 0)                  //如果位置是0,自己阻塞,等待其他线程来交换
                    return timed ?
                        awaitNanos(me, slot, nanos) :
                        await(me, slot);
                Object v = spinWait(me, slot);   //不是位置0,则自旋等待
                if (v != CANCEL)
                    return v;
                me = new Node(item);
            int m = max.get();
                if (m > (index >>>= 1))          //自旋的时候,没有线程来交换。执行下面的,index
                                                 //减半,挪个位置,重新开始for循环
        max.compareAndSet(m, m - 1);
            }
            else if (++fails > 1) {              //失败case1: slot有线程,要交互,但被其他
                                                 //线程抢了;case2: slot没有线程,自己准备占
                                                 //位置,又被其他线程抢了
                int m = max.get();
                if (fails > 3 && m < FULL && max.compareAndSet(m, m + 1))
                    index = m + 1;               //3次匹配失败,扩大index,再次开始for循环
                else if (--index < 0)
                    index = m;
            }
        }
    }
}
```

关于上面的代码,有几个点要说明:

(1)当一个线程调用 **exchange** 准备和其他线程交换数据的时候,无外乎两种情况。一种是

没有其他线程要交换数据,自己只能自旋或者阻塞,等待;另一种是恰好有其他线程在 Slot 里面等着,那么和对方交换。

(2)由于 Slot 不止 1 个,而是多个。如果运气好,根据自己的 thread id 找到对应的 Slot,里面恰好有别的线程在等待,就和对方交换。交换办法是:取出 Slot 指向的 Node,也就是对方的 Node,然后把这个 Node(本身也是一个 AtomicReference)指向自己的 item,唤醒对方,同时返回对方的 item。(再次强调一次:1 个 Node 实际上有 2 个 item 字段,1 个记录的是自己的 item,1 个记录的是对方的 item,因此实现 2 个线程的数据交换)。

(3)如果运气不好,Slot 是空的,如何处理呢?当前 Slot 为空,不代表其他 Slot 没有线程在等待。因此,如果当前 Slot 的 index = 0,自己就阻塞在那;如果 index ! = 0,则需要遍历所有的 Slot,看其他的 Slot 里面是否有线程在等待。最好是遍历一圈发现没有其他线程,自己再在 index = 0 的位置等待。

4.5 Phaser

4.5.1 用 Phaser 替代 CyclicBarrier 和 CountDownLatch

从 JDK7 开始,新增了一个同步工具类 Phaser,其功能比 CyclicBarrier 和 CountDownLatch 更加强大。

1. 用 Phaser 替代 CountDownLatch

考虑讲 CountDownLatch 时的例子,1 个主线程要等 10 个 worker 线程完成之后,才能做接下来的事情,也可以用 Phaser 来实现此功能。在 CountDownLatch 中,主要是 2 个函数:await() 和 countDown(),在 Phaser 中,与之相对应的函数是 awaitAdance(int n)和 arrive()。

```
Phaser ph = new Phaser(10);        //初始为 10
ph.awaitAdance(ph.getPhase());     //主线程调用该方法,阻塞在这。awaitAdance(int
                                   //phase),表示等待当前 phase(当前同步点)完成
ph.arrive();      //10 个 worker 线程,每个线程工作完成之后,调用 1 次 arrive()。顾名思义,
                  //"我到达了这个同步点"
```

2. 用 Phaser 替代 CyclicBarrier

考虑前面讲 CyclicBarrier 时,10 个工程师去公司应聘的例子,也可以用 Phaser 实现,代码基本类似。

```
Phaser ph = new Phaser(10);
run()
{
```

```
report("我已到公司");              //阶段1
ph.arriveAndAwaitAdvance();        //第1个同步点（代替CyclicBarrier的await）
笔试();                            //阶段2
ph.arriveAndAwaitAdvance();        //第2个同步点（代替CyclicBarrier的await）
面试();                            //阶段3
}
```

arriveAndAwaitAdance()就是 arrive()与 awaitAdvance(int)的组合，表示"我自己已到达这个同步点，同时要等待所有人都到达这个同步点，然后再一起前行"。

4.5.2 Phaser 新特性

特性 1：动态调整线程个数

CyclicBarrier 所要同步的线程个数是在构造函数中指定的，之后不能更改，而 Phaser 可以在运行期间动态地调整要同步的线程个数。Phaser 提供了下面这些函数来增加、减少所要同步的线程个数。

```
register()                      //注册1个
bulkRegister(int parties)       //注册多个
arriveAndDeregister()           //解注册
```

特性 2：层次 Phaser

多个 Phaser 可以组成如图 4-7 所示的树状结构，可以通过在构造函数中传入父 Phaser 来实现。

```
public Phaser(Phaser parent, int parties) {
    ...
}
```

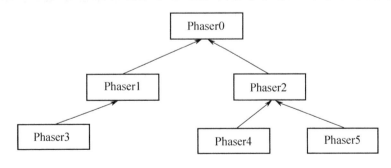

图 4-7　树状的 Phaser

先简单看一下 Phaser 内部关于树状结构的存储，如下面代码所示。

```
public class Phaser {
```

```
private final Phaser parent;
...
}
```

可以发现，在 Phaser 的内部结构中，每个 Phaser 记录了自己的父节点，但并没有记录自己的子节点列表。所以，每个 Phaser 知道自己的父节点是谁，但父节点并不知道自己有多少个子节点，对父节点的操作，是通过子节点来实现的。

树状的 Phaser 怎么使用呢？考虑如下代码，会组成如图 4-8 所示的树状 Phaser。

```
Phaser root = new Phaser(2);
Phaser c1 = new Phaser(root, 3);
Phaser c2 = new Phaser(root, 2);
Phaser c3 = new Phaser(c1,0);
```

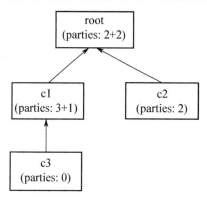

图 4-8 代码组成的树状 Phaser

本来 root 有两个参与者，然后为其加入了两个子 Phaser（c1，c2），每个子 Phaser 会算作 1 个参与者，root 的参与者就变成 2+2 = 4 个。c1 本来有 3 个参与者，为其加入了一个子 Phaser c3，参与者数量变成 3+1 = 4 个。c3 的参与者初始为 0，后续可以通过调用 register()函数加入。

对于树状 Phaser 上的每个节点来说，可以当作一个独立的 Phaser 来看待，其运作机制和一个单独的 Phaser 是一样的。具体来讲：父 Phaser 并不用感知子 Phaser 的存在，当子 Phaser 中注册的参与者数量大于 0 时，会把自己向父节点注册；当子 Phaser 中注册的参与者数量等于 0 时，会自动向父节点解注册。父 Phaser 把子 Phaser 当作一个正常参与的线程就可以了。

4.5.3 state 变量解析

大致了解了 Phaser 的用法和新特性之后，下面仔细剖析其实现原理。Phaser 没有基于 AQS 来实现，但具备 AQS 的核心特性：state 变量、CAS 操作、阻塞队列。先从 state 变量说起。

```
public class Phaser {
private volatile long state;
…
}
```

这个 64 位的 state 变量被拆成 4 部分,如图 4-9 所示为 state 变量各部分表示的意思。

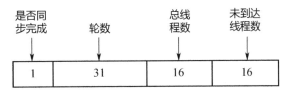

图 4-9 state 变量各部分表示的意思

最高位 0 表示未同步完成,1 表示同步完成,初始最高位为 0。

Phaser 提供了一系列的成员函数来从 state 中获取图 4-9 中的几个数字,如下所示。

```
//获取当前的轮数。当前轮数已经同步完成,返回值是一个负数(最高位为1)
public final int getPhase() {
return (int)(root.state >>> PHASE_SHIFT);   //当前 phase 未完成,返回值是一个负
                                            //数(最高位为1)
}
private static final int PHASE_SHIFT = 32;
//当前轮数同步完成,最高位为 1
public boolean isTerminated() {
return root.state < 0;
}
//获取总注册线程数
public int getRegisteredParties() {
return partiesOf(state);
}
private static int partiesOf(long s) {
return (int)s >>> PARTIES_SHIFT;    //先把 state 强制转成 32 位 int,再右移 16 位
}
private static final int PARTIES_SHIFT = 16;
//获取未到达的线程数
public int getUnarrivedParties() {
return unarrivedOf(reconcileState());
}
private static int unarrivedOf(long s) {
int counts = (int)s;
return (counts == EMPTY) ? 0 : counts & UNARRIVED_MASK;  //截取低 16 位
```

```
}
    private static final int  UNARRIVED_MASK = 0xffff;
```

下面再看一下 state 变量在构造函数中是如何被赋值的：

```
public Phaser(Phaser parent, int parties) {
    if (parties >>> PARTIES_SHIFT != 0)     //parties超出了最大个数（2的16次方），直
                                            //接抛出异常
        throw new IllegalArgumentException("Illegal number of parties");
    int phase = 0;       //初始轮数为0
    ...
    this.state = (parties == 0) ? (long)EMPTY :
    ((long)phase << PHASE_SHIFT) |      //或操作，赋给state。最高位为0，表示未同步完成
    ((long)parties << PARTIES_SHIFT) |
    ((long)parties);
}
    private static final int  EMPTY         = 1;
    private static final int  PARTIES_SHIFT = 16;
    private static final int  PHASE_SHIFT   = 32;
```

当 parties = 0 时，state 被赋予一个 EMPTY 常量，常量为 1；

当 parties != 0 时，把 phase 值左移 32 位；把 parties 左移 16 位；然后 parties 也作为最低的 16 位，3 个值做或操作，赋值给 state。这个赋值操作也反映了图 4-9 的意思。

4.5.4 阻塞与唤醒（Treiber Stack）

基于上述的 state 变量，对其执行 CAS 操作，并进行相应的阻塞与唤醒。如图 4-10 所示，右边的主线程会调用 awaitAdvance() 进行阻塞；左边的 arrive() 会对 state 进行 CAS 的累减操作，当未到达的线程数减到 0 时，唤醒右边阻塞的主线程。

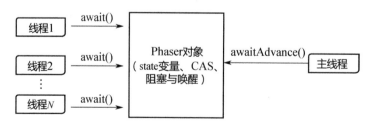

图 4-10　基于 state 的 CAS 的阻塞与唤醒示意图

在这里，阻塞使用的是一个称为 Treiber Stack 的数据结构，而不是 AQS 的双向链表。Treiber Stack 是一个无锁的栈，由 R. Kent Treiber 在其于 1986 年发表的论文 *Systems Programming: Coping with Parallelism* 中首次提出。它是一个单向链表，出栈、入栈都在链表头部，所以只需要一个

head 指针，而不需要 tail 指针。实现代码如下所示。

```
public class Phaser {
    static final class QNode implements ForkJoinPool.ManagedBlocker {
        QNode next;     //单向链表
        volatile Thread thread; //和 AQS 一样，每个 Node 对应一个 Thread 对象
        ...
    }
    private final AtomicReference<QNode> evenQ;    //链表的头指针
    private final AtomicReference<QNode> oddQ;
}
```

为了减少并发冲突，这里定义了 2 个链表，也就是 2 个 Treiber Stack。当 phase 为奇数轮的时候，阻塞线程放在 oddQ 里面；当 phase 为偶数轮的时候，阻塞线程放在 evenQ 里面。代码如下所示。

```
private AtomicReference<QNode> queueFor(int phase) {
    return ((phase & 1) == 0) ? evenQ : oddQ;
}
```

4.5.5　arrive()函数分析

下面看 arrive()函数是如何对 state 变量进行操作，又是如何唤醒线程的。

```
public int arrive() {
    return doArrive(false);
}
public int arriveAndDeregister() {
    return doArrive(true);
}
```

arrive()和 arriveAndDeregister()内部调用的都是 doArrive(boolean)函数。区别在于前者只是把"未达到线程数"减 1；后者则把"未到达线程数"和"下一轮的总线程数"都减 1。下面看一下 doArrive(boolean)函数的实现。

```
private int doArrive(boolean deregister) {
    int adj = deregister ? ONE_ARRIVAL|ONE_PARTY : ONE_ARRIVAL;
                                    //最低 16 位减 1 或者低 32 位中的 2 个 16 位都减 1
    final Phaser root = this.root;
    for (;;) {
        long s = (root == this) ? state : reconcileState();
        int phase = (int)(s >>> PHASE_SHIFT);
        int counts = (int)s;
        int unarrived = (counts & UNARRIVED_MASK) - 1;
```

```
        if (phase < 0)
            return phase;
        else if (counts == EMPTY || unarrived < 0) {
            if (root == this || reconcileState() == s)
                throw new IllegalStateException(badArrive(s));
        }
        else if (UNSAFE.compareAndSwapLong(this, stateOffset, s, s-=adj)) { //CAS减1
            if (unarrived == 0) {   //所有线程到达
                long n = s & PARTIES_MASK;
                int nextUnarrived = (int)n >>> PARTIES_SHIFT;
                if (root != this)   //如果父Phaser不为空
                    return parent.doArrive(nextUnarrived == 0); //先调用parent的doArrive
                if (onAdvance(phase, nextUnarrived))
                    n |= TERMINATION_BIT;    //最高位置为1
                else if (nextUnarrived == 0)
                    n |= EMPTY;
                else
                    n |= nextUnarrived;   //下一轮的未到达数等于总的线程个数
                n |= (long)((phase + 1) & MAX_PHASE) << PHASE_SHIFT; //phase加1
                UNSAFE.compareAndSwapLong(this, stateOffset, s, n);  //重置state
                releaseWaiters(phase);  //唤醒队列中等待的线程
            }
            return phase;    //如果所有线程没有到达，只是把未到达线程数减1，什么都不做，该
                             //函数返回
        }
    }
}
```

关于上面的函数，有以下几点说明：

（1）定义了 2 个常量如下。当 deregister = false 时，只最低的 16 位需要减 1，adj = ONE_ARRIVAL；当 deregister = true 时，低 32 位中的 2 个 16 位都需要减 1，adj = ONE_ARRIVAL| ONE_PARTY。

```
    private static final int  ONE_ARRIVAL     = 1;
    private static final int  ONE_PARTY       = 1 << PARTIES_SHIFT;
```

（2）把未到达线程数减 1。减了之后，如果还未到 0，什么都不做，直接返回。如果到 0，会做 2 件事情：第 1，重置 state，把 state 的未到达线程个数重置到总的注册的线程数中，同时 phase 加 1；第 2，唤醒队列中的线程。

下面看一下唤醒函数：

```
    private void releaseWaiters(int phase) {
```

```
QNode q;
Thread t;
AtomicReference<QNode> head = (phase & 1) == 0 ? evenQ : oddQ;
                                //根据phase是奇数还是偶数，决定用evenQ还是oddQ。
while ((q = head.get()) != null &&
    q.phase != (int)(root.state >>> PHASE_SHIFT)) {   //遍历栈
  if (head.compareAndSet(q, q.next) &&
    (t = q.thread) != null) {
    q.thread = null;
    LockSupport.unpark(t);
  }
 }
}
```

遍历整个栈，只要栈当中节点的 phase 不等于当前 Phaser 的 phase，说明该节点不是当前轮的，而是前一轮的，应该被释放并唤醒。

接下来看一下线程是如何被阻塞的。

4.5.6　awaitAdvance()函数分析

```
public int awaitAdvance(int phase) {
final Phaser root = this.root;
long s = (root == this) ? state : reconcileState();  //当只有一个Phaser，没
                                //有树状结构时, root=this
int p = (int)(s >>> PHASE_SHIFT);
if (phase < 0)     //phase已经结束，不用阻塞了，直接返回
    return phase;
if (p == phase)
    return root.internalAwaitAdvance(phase, null);   //阻塞在phase这一轮上面
return p;
}
```

下面的 while 循环中有 4 个分支：初始的时候，node == null，进入第 1 个分支进行自旋，自旋次数满足之后，会新建一个 QNode 节点；之后执行第 3、第 4 个分支，分别把该节点入栈并阻塞。

```
private int internalAwaitAdvance(int phase, QNode node) {
releaseWaiters(phase-1);
boolean queued = false;
int lastUnarrived = 0;
int spins = SPINS_PER_ARRIVAL;
long s;
int p;
```

```
        while ((p = (int)((s = state) >>> PHASE_SHIFT)) == phase) {
            if (node == null) {              //自旋
                int unarrived = (int)s & UNARRIVED_MASK;
                if (unarrived != lastUnarrived &&
                    (lastUnarrived = unarrived) < NCPU)
                    spins += SPINS_PER_ARRIVAL;
                boolean interrupted = Thread.interrupted();
                if (interrupted || --spins < 0) {  //由自旋结束,建一个节点,接下来进入阻塞
                    node = new QNode(this, phase, false, false, 0L);
                    node.wasInterrupted = interrupted;
                }
            }
            else if (node.isReleasable())    //由阻塞中唤醒,退出 while 循环
                break;
            else if (!queued) {
                AtomicReference<QNode> head = (phase & 1) == 0 ? evenQ : oddQ;
                QNode q = node.next = head.get();
                if ((q == null || q.phase == phase) &&
                    (int)(state >>> PHASE_SHIFT) == phase)
                    queued = head.compareAndSet(q, node);    //节点入栈
            }
            else {
                try {
                    ForkJoinPool.managedBlock(node);    //调用 node.block()阻塞
                } catch (InterruptedException ie) {
                    node.wasInterrupted = true;
                }
            }
        }
```

这里调用了 ForkJoinPool.managedBlock(ManagedBlocker blocker)函数,目的是把 node 对应的线程阻塞。ManagerdBlocker 是 ForkJoinPool 里面的一个接口, 定义如下:

```
public static interface ManagedBlocker {
    boolean block() throws InterruptedException;
    boolean isReleasable();
}
```

QNode 实现了该接口,实现原理还是 park(),如下所示。之所以没有直接使用 park()/unpark()来实现阻塞、唤醒,而是封装了 ManagedBlocker 这一层,主要是出于使用上的方便考虑。一方面是 park()可能被中断唤醒,另一方面是带超时时间的 park(),把这二者都封装在一起。

```
static final class QNode implements ForkJoinPool.ManagedBlocker
    public boolean isReleasable() {
```

```
    if (thread == null)
        return true;
    if (phaser.getPhase() != phase) {
        thread = null;
        return true;
    }
    if (Thread.interrupted())
        wasInterrupted = true;
    if (wasInterrupted && interruptible) {
        thread = null;
        return true;
    }
    if (timed) {
        if (nanos > 0L) {
            long now = System.nanoTime();
            nanos -= now - lastTime;
            lastTime = now;
        }
        if (nanos <= 0L) {
            thread = null;
            return true;
        }
    }
    return false;
}

public boolean block() {    //block 函数封装了 park()和 parkNanos(xx), 使用起来更方便
    if (isReleasable())
        return true;
    else if (!timed)
        LockSupport.park(this);
    else if (nanos > 0)
        LockSupport.parkNanos(this, nanos);
    return isReleasable();
}
...
}
```

理解了 arrive()和 awaitAdvance(), arriveAndAwaitAdvance()就是二者的一个组合版本, 此处不再展开分析。

第 5 章 并发容器

在 Lock 和 Phaser 的实现中,已经介绍了基于 CAS 实现的无锁队列和无锁栈。本章将全面介绍 Concurrent 包提供的各种并发容器。

5.1 BlockingQueue

在所有的并发容器中,BlockingQueue 是最常见的一种。BlockingQueue 是一个带阻塞功能的队列,当入队列时,若队列已满,则阻塞调用者;当出队列时,若队列为空,则阻塞调用者。

在 Concurrent 包中,BlockingQueue 是一个接口,有许多个不同的实现类,如图 5-1 所示。

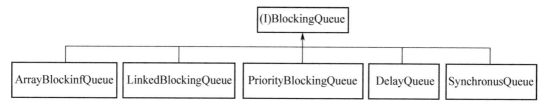

图 5-1 BlockingQueue 的各种实现类

该接口的定义如下:

```
public interface BlockingQueue<E> extends Queue<E> {
...
boolean add(E e);
boolean offer(E e);
void put(E e) throws InterruptedException;
```

```
E take() throws InterruptedException;
...
}
```

可以看到，该接口和 JDK 集合包中的 Queue 接口是兼容的，同时在其基础上增加了阻塞功能。在这里，入队提供了 add(..)、offer(..)、put (..)3 个函数，有什么区别呢？从上面的定义可以看到，add(..)和 offer(..)的返回值是布尔类型，而 put 无返回值，还会抛出中断异常，所以 add(..)和 offer(..)是无阻塞的，也是 Queue 本身定义的接口，而 put(..)是阻塞式的。add(..)和 offer(..)的区别不大，当队列为满的时候，前者会抛出异常，后者则直接返回 false。

出队列与之类似，提供了 remove()、peek()、take()等函数，remove()和 peek()是非阻塞式的，take()是阻塞式的。

下面分别介绍 BlockingQueue 的各种不同实现。

5.1.1 ArrayBlockingQueue

ArrayBlockingQueue 是一个用数组实现的环形队列，在构造函数中，会要求传入数组的容量。

```
public ArrayBlockingQueue(int capacity, boolean fair) {
...
}
```

其核心数据结构如下：

```
public class ArrayBlockingQueue<E> extends AbstractQueue<E>
        implements BlockingQueue<E>, java.io.Serializable {
...
    final Object[] items;    //数组及队头、队尾指针
    int takeIndex;
    int putIndex;
    int count;
    //其核心就是1把锁 + 2个条件
    final ReentrantLock lock;
    private final Condition notEmpty;
    private final Condition notFull;
...
}
```

其 put/take 函数也很简单，如下所示。

```
public void put(E e) throws InterruptedException {
    checkNotNull(e);
```

```java
        final ReentrantLock lock = this.lock;
        lock.lockInterruptibly();          //使用的可中断的Lock
        try {
            while (count == items.length)
                notFull.await();           //put的时候,若队列满了,则阻塞
            insert(e);
        } finally {
            lock.unlock();
        }
    }
    private void insert(E x) {
        items[putIndex] = x;
        putIndex = inc(putIndex);
        ++count;
        notEmpty.signal();   //当put进去之后,通知非空条件
    }
    public E take() throws InterruptedException {
        final ReentrantLock lock = this.lock;
        lock.lockInterruptibly();
        try {
            while (count == 0)
                notEmpty.await(); //在take的时候,若队列为空,则阻塞
            return extract();
        } finally {
            lock.unlock();
        }
    }
    private E extract() {
        final Object[] items = this.items;
        E x = this.<E>cast(items[takeIndex]);
        items[takeIndex] = null;
        takeIndex = inc(takeIndex);
        --count;
        notFull.signal();    //take结束,通知非满条件
        return x;
    }
}
```

5.1.2 LinkedBlockingQueue

LinkedBlockingQueue 是一种基于单向链表的阻塞队列。因为队头和队尾是 2 个指针分开操作的,所以用了 2 把锁 + 2 个条件,同时有 1 个 AtomicInteger 的原子变量记录 count 数。

```java
public class LinkedBlockingQueue<E> extends AbstractQueue<E>
    implements BlockingQueue<E>, java.io.Serializable {
...
    private final int capacity;
private final AtomicInteger count = new AtomicInteger(0);   //原子变量
    private transient Node<E> head;       //单向链表的头部和尾部
    private transient Node<E> last;
    //2 把锁 + 2 个条件
    private final ReentrantLock takeLock = new ReentrantLock();
    private final Condition notEmpty = takeLock.newCondition();
    private final ReentrantLock putLock = new ReentrantLock();
    private final Condition notFull = putLock.newCondition();
...
}
```

在其构造函数中,也可以指定队列的总容量。如果不指定,默认为 Integer.MAX_VALUE。

```java
public LinkedBlockingQueue() {
    this(Integer.MAX_VALUE);
}

public LinkedBlockingQueue(int capacity) {
    if (capacity <= 0) throw new IllegalArgumentException();
    this.capacity = capacity;
    last = head = new Node<E>(null);
}
```

下面看一下其 put/take 实现。

```java
public void put(E e) throws InterruptedException {
    if (e == null) throw new NullPointerException();
    int c = -1;
    Node<E> node = new Node(e);
    final ReentrantLock putLock = this.putLock;
    final AtomicInteger count = this.count;
    putLock.lockInterruptibly();
    try {
        while (count.get() == capacity) {
            notFull.await();
        }
        enqueue(node);
        c = count.getAndIncrement();
        if (c + 1 < capacity)
            notFull.signal();     //通知其他 put 线程
    } finally {
        putLock.unlock();
```

```java
        }
        if (c == 0)
            signalNotEmpty();    //这个里面要加takeLock
}

public E take() throws InterruptedException {
    E x;
    int c = -1;
    final AtomicInteger count = this.count;
    final ReentrantLock takeLock = this.takeLock;
    takeLock.lockInterruptibly();
    try {
        while (count.get() == 0) {
            notEmpty.await();
        }
        x = dequeue();
        c = count.getAndDecrement();
        if (c > 1)
            notEmpty.signal();      //通知其他take线程
    } finally {
        takeLock.unlock();
    }
    if (c == capacity)
        signalNotFull();            //这个里面要加putLock
    return x;
}
```

LinkedBlockingQueue 和 ArrayBlockingQueue 的实现有一些差异, 有几点要特别说明:

(1) 为了提高并发度,用2把锁,分别控制队头、队尾的操作。意味着在 put(..)和 put(..)之间、take()与 take()之间是互斥的, put(..)和 take()之间并不互斥。但对于 count 变量, 双方都需要操作,所以必须是原子类型。

(2) 因为各自拿了一把锁, 所以当需要调用对方的 condition 的 signal 时, 还必须再加上对方的锁, 就是 signalNotEmpty()和 signalNotFull()函数。示例如下所示。

```java
private void signalNotEmpty() {
    final ReentrantLock takeLock = this.takeLock;
    takeLock.lock();        //必须先拿到takeLock,才能调用notEmpty.signal
    try {
        notEmpty.signal();
    } finally {
        takeLock.unlock();
    }
}
```

```
}
    private void signalNotFull() {
    final ReentrantLock putLock = this.putLock;
    putLock.lock();    //必须先拿到putLock，才能调用notFull.signal
    try {
        notFull.signal();
    } finally {
        putLock.unlock();
    }
}
```

（3）不仅 put 会通知 take，take 也会通知 put。当 put 发现非满的时候，也会通知其他 put 线程；当 take 发现非空的时候，也会通知其他 take 线程。

5.1.3 PriorityBlockingQueue

队列通常是先进先出的，而 PriorityQueue 是按照元素的优先级从小到大出队列的。正因为如此，PriorityQueue 中的 2 个元素之间需要可以比较大小，并实现 Comparable 接口。

其核心数据结构如下：

```
    public class PriorityBlockingQueue<E> extends AbstractQueue<E>
    implements BlockingQueue<E>, java.io.Serializable {
    ...
    private transient Object[] queue;   //用数组实现的二叉小根堆
private transient int size;

private transient Comparator<? super E> comparator;
//1把锁 + 1个条件，没有非满的条件
    private final ReentrantLock lock;
private final Condition notEmpty;
    ...
    }
```

其构造函数如下所示，如果不指定初始大小，内部会设定一个默认值 11，当元素个数超过这个大小之后，会自动扩容。

```
    public PriorityBlockingQueue() {
    this(DEFAULT_INITIAL_CAPACITY, null);
}
    private static final int DEFAULT_INITIAL_CAPACITY = 11;
    public PriorityBlockingQueue(int initialCapacity,
                    Comparator<? super E> comparator) {
    if (initialCapacity < 1)
```

```
        throw new IllegalArgumentException();
    this.lock = new ReentrantLock();
    this.notEmpty = lock.newCondition();
    this.comparator = comparator;
    this.queue = new Object[initialCapacity];
}
```

下面是对应的 put/take 函数的实现。

```
    public void put(E e) {
    offer(e);
}

    public boolean offer(E e) {
    if (e == null)
        throw new NullPointerException();
    final ReentrantLock lock = this.lock;
    lock.lock();
    int n, cap;
    Object[] array;
    while ((n = size) >= (cap = (array = queue).length))
        tryGrow(array, cap);       //size 超出了数组的长度,扩容
    try {
        Comparator<? super E> cmp = comparator;
        if (cmp == null)           //没有定义比较操作符,使用元素自带的比较功能
            siftUpComparable(n, e, array);    //元素入堆,也就是执行 siftUp 操作
        else
            siftUpUsingComparator(n, e, array, cmp);
        size = n + 1;
        notEmpty.signal();
    } finally {
        lock.unlock();
    }
    return true;
}
    public E take() throws InterruptedException {
    final ReentrantLock lock = this.lock;
    lock.lockInterruptibly();
    E result;
    try {
        while ( (result = dequeue()) == null)     //出队列
            notEmpty.await();
    } finally {
        lock.unlock();
```

```
        }
        return result;
    }

    private E dequeue() {
        int n = size - 1;
        if (n < 0)
            return null;
        else {
            Object[] array = queue;
            E result = (E) array[0];   //因为是最小二叉堆,堆顶即是要出队的元素
            E x = (E) array[n];
            array[n] = null;
            Comparator<? super E> cmp = comparator;
            if (cmp == null)
                siftDownComparable(0, x, array, n);//调整堆,也就是执行siftDown操作
            else
                siftDownUsingComparator(0, x, array, n, cmp);
            size = n;
            return result;
        }
    }
}
```

从上面可以看到,在阻塞的实现方面,和 ArrayBlockingQueue 的机制相似,主要区别是用数组实现了一个二叉堆,从而实现按优先级从小到大出队列。另一个区别是没有 notFull 条件,当元素个数超出数组长度时,执行扩容操作。

5.1.4 DelayQueue

DelayQueue 即延迟队列,也就是一个按延迟时间从小到大出队的 PriorityQueue。所谓延迟时间,就是"未来将要执行的时间"-"当前时间"。为此,放入 DelayQueue 中的元素,必须实现 Delayed 接口,如下所示。

```
public interface Delayed extends Comparable<Delayed> {
    long getDelay(TimeUnit unit);
}
```

关于该接口,有两点说明:

(1)如果 getDelay 的返回值小于或等于 0,则说明该元素到期,需要从队列中拿出来执行。

(2)该接口首先继承了 Comparable 接口,所以要实现该接口,必须实现 Comparable 接口。具体来说,就是基于 getDelay() 的返回值比较两个元素的大小。

下面看一下 DelayQueue 的核心数据结构。

```java
public class DelayQueue<E extends Delayed> extends AbstractQueue<E>
    implements BlockingQueue<E> {
...
private final PriorityQueue<E> q = new PriorityQueue<E>();   //优先级队列
//1把锁 + 1个非空条件
private final transient ReentrantLock lock = new ReentrantLock();
private final Condition available = lock.newCondition();
...
}
```

下面介绍 put/take 的实现，先从 take 说起，因为这样更能看出 DelayQueue 的特性。

```java
public E take() throws InterruptedException {
    final ReentrantLock lock = this.lock;
    lock.lockInterruptibly();
    try {
        for (;;) {
            E first = q.peek();          //取出二叉堆的堆顶元素，也就是延迟时间最小的
            if (first == null)
                available.await();       //队列为空，take 线程阻塞
            else {
                long delay = first.getDelay(NANOSECONDS);
                if (delay <= 0)          //堆顶元素的延迟时间小于或等于0，出队列，返回
                    return q.poll();
                first = null;
                if (leader != null)      //如果已经有其他线程也在等待这个元素，则无限期阻塞
                    available.await();
                else {
                    Thread thisThread = Thread.currentThread();
                    leader = thisThread;
                    try {
                        available.awaitNanos(delay);    //否则阻塞有限的时间
                    } finally {
                        if (leader == thisThread)
                            leader = null;
                    }
                }
            }
        }
    } finally {
        if (leader == null && q.peek() != null)
            available.signal();          //自己是 leader，已经获取了堆顶元素，唤醒其他线程
        lock.unlock();
```

        }
}
关于take()函数,有两点需要说明:

(1)不同于一般的阻塞队列,只在队列为空的时候,才阻塞。如果堆顶元素的延迟时间没到,也会阻塞。

(2)在上面的代码中使用了一个优化技术,用一个Thread leader变量记录了等待堆顶元素的第 1 个线程。为什么这样做呢?通过getDelay(..)可以知道堆顶元素何时到期,不必无限期等待,可以使用condition.awaitNanos()等待一个有限的时间;只有当发现还有其他线程也在等待堆顶元素(leader != NULL)时,才需要无限期等待。

下面看一下 put 的实现。

```
public void put(E e) {
    offer(e);
}
public boolean offer(E e) {
    final ReentrantLock lock = this.lock;
    lock.lock();
    try {
        q.offer(e);    //元素放入二叉堆
        if (q.peek() == e) { //如果放进去的元素刚好在堆顶,说明放入的元素延迟时间是最小
                             //的,那么需要通知等待的线程,否则放入的元素不在堆顶,没有
                             //必要通知等待的线程
            leader = null;
            available.signal();
        }
        return true;
    } finally {
        lock.unlock();
    }
}
```

关于上面的实现,有一点要说明:不是每放入一个元素,都需要通知等待的线程。放入的元素,如果其延迟时间大于当前堆顶的元素延迟时间,就没必要通知等待的线程;只有当延迟时间是最小的,在堆顶时,才有必要通知等待的线程,也就是上面代码中的 if(q.peek() == e)段落。

5.1.5 SynchronousQueue

SynchronousQueue 是一种特殊的 BlockingQueue,它本身没有容量。先调 put(..),线程会阻塞;直到另外一个线程调用了 take(),两个线程才同时解锁,反之亦然。对于多个线程而言,例

如 3 个线程，调用 3 次 put(..)，3 个线程都会阻塞；直到另外的线程调用 3 次 take()，6 个线程才同时解锁，反之亦然。

在讲解线程池中的 CachedThreadPool 实现的时候会使用到 SynchronousQueue 的这种特性。

接下来就看一下，SynchronousQueue 是如何实现的。先从构造函数看起。

```
public SynchronousQueue(boolean fair) {
    transferer = fair ? new TransferQueue() : new TransferStack();
}
```

和锁一样，也有公平和非公平模式。如果是公平模式，则用 TransferQueue 实现；如果是非公平模式，则用 TransferStack 实现。这两个类分别是什么呢？先看一下 put/take 的实现。

```
public void put(E o) throws InterruptedException {
    if (o == null) throw new NullPointerException();
    if (transferer.transfer(o, false, 0) == null) {
        Thread.interrupted();
        throw new InterruptedException();
    }
}

public E take() throws InterruptedException {
    Object e = transferer.transfer(null, false, 0);
    if (e != null)
        return (E)e;
    Thread.interrupted();
    throw new InterruptedException();
}
```

可以看到，put/take 都调用了 transfer(..)接口。而 TransferQueue 和 TransferStack 分别实现了这个接口。该接口在 SynchronousQueue 内部，如下所示。如果是 put(..)，则第 1 个参数就是对应的元素；如果是 take()，则第 1 个参数为 null。后 2 个参数分别为是否设置超时和对应的超时时间。

```
abstract static class Transferer {
    abstract Object transfer(Object e, boolean timed, long nanos);
}
```

接下来看一下什么是公平模式和非公平模式，如图 5-2 所示。假设 3 个线程分别调用了 put(..)，3 个线程会进入阻塞状态，直到其他线程调用 3 次 take()，和 3 个 put(..)——配对。

如果是公平模式（队列模式），则第 1 个调用 put(..)的线程 1 会在队列头部，第 1 个到来的 take()线程和它进行配对，遵循先到先配对的原则，所以是公平的；如果是非公平模式（栈模式），则第 3 个调用 put(..)的线程 3 会在栈顶，第 1 个到来的 take()线程和它进行配对，遵循的是后到先配对的原则,所以是非公平的。

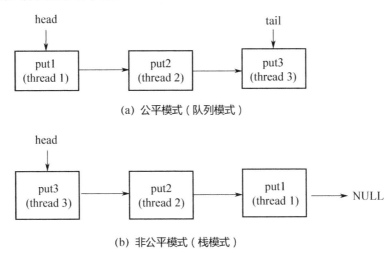

(a) 公平模式(队列模式)

(b) 非公平模式(栈模式)

图 5-2 公平模式与非公平模式对比

下面分别看一下 TransferQueue 和 TransferStack 的实现。

1. TransferQueue

```
public class SynchronousQueue<E> extends AbstractQueue<E>
implements BlockingQueue<E>, java.io.Serializable {
…
static final class TransferQueue extends Transferer {
static final class QNode {
volatile QNode next;            //单向链表
volatile Object item;           //如果是put,则item!=null;如果是take,则item=null
volatile Thread waiter;         //put/take 对应的阻塞线程和上面的 item 对应,
final boolean isData;           //put, isData = true,否则为 false
…
}
transient volatile QNode head;  //单向链表的头和尾
transient volatile QNode tail;
…
}
}
```

从上面的代码可以看出,TransferQueue 是一个基于单向链表而实现的队列,通过 head 和 tail 2 个指针记录头部和尾部。初始的时候,head 和 tail 会指向一个空节点,构造函数如下所示。

```
TransferQueue() {
```

```
QNode h = new QNode(null, false);   //空节点
head = h;
tail = h;
}
```

图 5-3 所示为 TransferQueue 的工作原理。

阶段（a）：队列中是一个空的节点，head/tail 都指向这个空节点。

阶段（b）：3 个线程分别调用 put，生成 3 个 QNode，进入队列。

阶段（c）：来了一个线程调用 take，会和队列头部的第 1 个 QNode 进行配对。

阶段（d）：第 1 个 QNode 出队列。

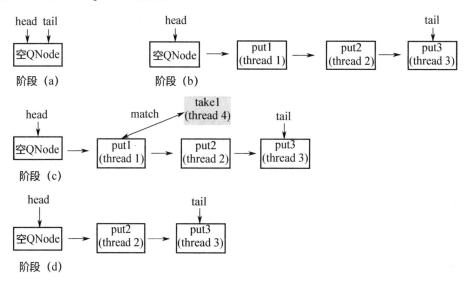

图 5-3 TransferQueue 的工作原理

这里有一个关键点：put 节点和 take 节点一旦相遇，就会配对出队列，所以在队列中不可能同时存在 put 节点和 take 节点，要么所有节点都是 put 节点，要么所有节点都是 take 节点。

接下来看一下 TransferQueue 的代码实现。

```
Object transfer(Object e, boolean timed, long nanos) {
QNode s = null;
boolean isData = (e != null);

for (;;) {
    QNode t = tail;
    QNode h = head;
```

```
    if (t == null || h == null) //队列还未初始化，自旋等待
        continue;

    if (h == t || t.isData == isData) { //队列为空或者当前线程和队列中元素为同一
                                        //种模式
        QNode tn = t.next;
        if (t != tail)            //不一致读，重新执行 for 循环
            continue;
        if (tn != null) {
            advanceTail(t, tn);
            continue;
        }
        if (timed && nanos <= 0)
            return null;
        if (s == null)
            s = new QNode(e, isData);      //新建一个节点
        if (!t.casNext(null, s))           //加入尾部
            continue;

        advanceTail(t, s);                 //后移 tail 指针
        Object x = awaitFulfill(s, e, timed, nanos);   //进入阻塞状态
        if (x == s) {
            clean(t, s);
            return null;
        }

        if (!s.isOffList()) {   //从阻塞中唤醒，确定已经处于队列中的第 1 个元素
            advanceHead(t, s);
            if (x != null)
                s.item = s;
            s.waiter = null;
        }
        return (x != null) ? x : e;

    } else {        //当前线程可以和队列中的第 1 个元素进行配对
        QNode m = h.next;              //取队列中第 1 个元素
        if (t != tail || m == null || h != head)   //不一致读，重新执行 for 循环
            continue;

        Object x = m.item;
        if (isData == (x != null) ||      // 已经配对过
            x == m ||
```

```
                !m.casItem(x, e)) {           //尝试配对
                advanceHead(h, m);            //已经配对过，直接出队列
                continue;
            }

            advanceHead(h, m);                //配对成功，出队列
            LockSupport.unpark(m.waiter);     //唤醒队列中与第 1 个元素对应的线程
            return (x != null) ? x : e;
        }
    }
}
```

整个 for 循环有两个大的 if-else 分支，如果当前线程和队列中的元素是同一种模式（都是 put 节点或者 take 节点），则与当前线程对应的节点被加入队列尾部并且阻塞；如果不是同一种模式，则选取队列头部的第 1 个元素进行配对。

这里的配对就是 m.casItem(x, e)，把自己的 item x 换成对方的 item e，如果 CAS 操作成功，则配对成功。如果是 put 节点，则 isData = true，item != null；如果是 take 节点，则 isData = false，item = null。如果 CAS 操作不成功，则 isData 和 item 之间将不一致，也就是 isData != (x != null)，通过这个条件可以判断节点是否已经被匹配过了。

2. TransferStack

TransferStack 的定义如下所示，首先，它也是一个单向链表。不同于队列，只需要 head 指针就能实现入栈和出栈操作。

链表中的节点有三种状态，REQUEST 对应 take 节点，DATA 对应 put 节点，二者配对之后，会生成一个 FULFILLING 节点，入栈，然后 FULLING 节点和被配对的节点一起出栈。

```
    static final class TransferStack extends Transferer {
    static final int REQUEST   = 0;
static final int DATA       = 1;
static final int FULFILLING = 2;
    static final class SNode {
    volatile SNode next;            //单向链表
        volatile SNode match;       //配对的节点
    volatile Thread waiter;         //对应的阻塞线程
    Object item;
    int mode;                       //三种模式
    ...}
    volatile SNode head;
    }
```

如图 5-4 所示为 TransferStack 的工作原理。

阶段（a）：head 指向 NULL。不同于 TransferQueue，这里没有空的头节点。

阶段（b）：3 个线程调用 3 次 put，依次入栈。

阶段（c）：线程 4 调用 take，和栈顶的第 1 个元素配对，生成 FULLFILLING 节点，入栈。

阶段（d）：栈顶的 2 个元素同时入栈。

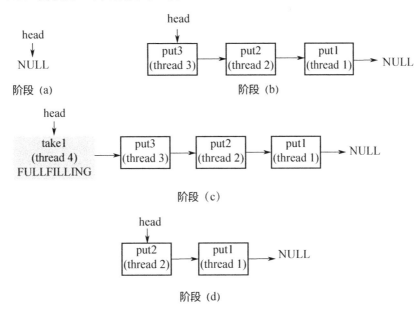

图 5-4　TransferStack 的工作原理

下面看一下具体的代码实现。

```
Object transfer(Object e, boolean timed, long nanos) {
SNode s = null; // constructed/reused as needed
int mode = (e == null) ? REQUEST : DATA;

for (;;) {
   SNode h = head;
   if (h == null || h.mode == mode) {   //同一种模式
      if (timed && nanos <= 0) {
         if (h != null && h.isCancelled())
            casHead(h, h.next);
         else
            return null;
```

```java
    } else if (casHead(h, s = snode(s, e, h, mode))) {    //入栈
        SNode m = awaitFulfill(s, timed, nanos);    //阻塞等待
        if (m == s) {
            clean(s);
            return null;
        }
        if ((h = head) != null && h.next == s)
            casHead(h, s.next);
        return (mode == REQUEST) ? m.item : s.item;
    }
} else if (!isFulfilling(h.mode)) {    //非同一种模式，待匹配
    if (h.isCancelled())
        casHead(h, h.next);
    else if (casHead(h, s=snode(s, e, h, FULFILLING|mode))) {
        //生成一个FULFILLING节点，入栈
        for (;;) {
            SNode m = s.next;
            if (m == null) {
                casHead(s, null);
                s = null;
                break;
            }
            SNode mn = m.next;
            if (m.tryMatch(s)) {
                casHead(s, mn);    //两个节点一起出栈
                return (mode == REQUEST) ? m.item : s.item;
            } else
                s.casNext(m, mn);
        }
    }
} else {                                  //已经匹配过了，出栈
    SNode m = h.next;
    if (m == null)
        casHead(h, null);
    else {
        SNode mn = m.next;
        if (m.tryMatch(h))
            casHead(h, mn);            //配对，一起出栈
        else
            h.casNext(m, mn);
    }
}
```

        }
    }

5.2 BlockingDeque

BlockingDeque 定义了一个阻塞的双端队列接口，如下所示。

```
public interface BlockingDeque<E> extends BlockingQueue<E>, Deque<E> {
void putFirst(E e) throws InterruptedException;
void putLast(E e) throws InterruptedException;
E takeFirst() throws InterruptedException;
E takeLast() throws InterruptedException;
…
}
```

可以看到，该接口在继承了 BlockingQueue 接口的同时，增加了对应的双端队列操作接口。该接口只有一个实现，就是 LinkedBlockingDeque。

其核心数据结构如下所示，是一个双向链表。

```
public class LinkedBlockingDeque<E> extends AbstractQueue<E>
    implements BlockingDeque<E>, java.io.Serializable {
    static final class Node<E> {
        E item;
        Node<E> prev;              //双向链表的 Node
        Node<E> next;
        Node(E x) {
            item = x;
        }
    }
    transient Node<E> first;           //队列的头和尾
    transient Node<E> last;
    private transient int count;       //元素个数
    private final int capacity;        //容量
    //1 把锁 + 2 个条件
    final ReentrantLock lock = new ReentrantLock();
    private final Condition notEmpty = lock.newCondition();
    private final Condition notFull = lock.newCondition();
    …
}
```

对应的实现原理，和 LinkedBlockingQueue 基本一样，只是 LinkedBlockingQueue 是单向链表，而 LinkedBlockingDeque 是双向链表。

```java
public E takeFirst() throws InterruptedException {
    final ReentrantLock lock = this.lock;
    lock.lock();
    try {
        E x;
        while ( (x = unlinkFirst()) == null)
            notEmpty.await();
        return x;
    } finally {
        lock.unlock();
    }
}

public E takeLast() throws InterruptedException {
    final ReentrantLock lock = this.lock;
    lock.lock();
    try {
        E x;
        while ( (x = unlinkLast()) == null)
            notEmpty.await();
        return x;
    } finally {
        lock.unlock();
    }
}

public void putFirst(E e) throws InterruptedException {
    if (e == null) throw new NullPointerException();
    Node<E> node = new Node<E>(e);
    final ReentrantLock lock = this.lock;
    lock.lock();
    try {
        while (!linkFirst(node))
            notFull.await();
    } finally {
        lock.unlock();
    }
}

public void putLast(E e) throws InterruptedException {
    if (e == null) throw new NullPointerException();
    Node<E> node = new Node<E>(e);
    final ReentrantLock lock = this.lock;
    lock.lock();
    try {
```

```
        while (!linkLast(node))
            notFull.await();
    } finally {
        lock.unlock();
    }
}
```

5.3 CopyOnWrite

CopyOnWrite 指在"写"的时候,不是直接"写"源数据,而是把数据拷贝一份进行修改,再通过悲观锁或者乐观锁的方式写回。那为什么不直接修改,而是要拷贝一份修改呢?这是为了在"读"的时候不加锁。下面通过几个案例来展现 CopyOnWrite 的应用。

5.3.1 CopyOnWriteArrayList

和 ArrayList 一样,CopyOnWriteArrayList 的核心数据结构也是一个数组,代码如下。

```
public class CopyOnWriteArrayList<E>
implements List<E>, RandomAccess, Cloneable, java.io.Serializable {
...
private volatile transient Object[] array;
}
```

下面是 CopyOnArrayList 的几个"读"函数:

```
final Object[] getArray() {
        return array;
    }
    public E get(int index) {                       //未加锁
        return (E)(getArray()[index]);
    }
    public boolean isEmpty() {
        return size() == 0;
    }
    public boolean contains(Object o) {
        Object[] elements = getArray();
        return indexOf(o, elements, 0, elements.length) >= 0;
    }
    public int indexOf(Object o) {
        Object[] elements = getArray();
        return indexOf(o, elements, 0, elements.length);
    }
```

```java
    private static int indexOf(Object o, Object[] elements,
                    int index, int fence) {
        if (o == null) {
            for (int i = index; i < fence; i++)
                if (elements[i] == null)
                    return i;
        } else {
            for (int i = index; i < fence; i++)
                if (o.equals(elements[i]))
                    return i;
        }
        return -1;
    }
```

既然这些"读"函数都没有加锁,那么是如何保证"线程安全"呢?答案在"写"函数里面。

```java
public boolean add(E e) {
    final ReentrantLock lock = this.lock;
    lock.lock();        //加悲观锁
    try {
        Object[] elements = getArray();
        int len = elements.length;
        Object[]newElements = Arrays.copyOf(elements, len + 1);
        //CopyOnWrite,写的时候,先拷贝一份之前的数组
        newElements[len] = e;
        setArray(newElements);    //把新数组赋值给老数组
        return true;
    } finally {
        lock.unlock();
    }
}

    final void setArray(Object[] a) {
        array = a;
    }
```

其他"写"函数,例如 remove 和 add 类似,此处不再详述。

5.3.2 CopyOnWriteArraySet

CopyOnWriteArraySet 就是用 Array 实现的一个 Set,保证所有元素都不重复。其内部是封装的一个 CopyOnWriteArrayList。

```
public class CopyOnWriteArraySet<E> extends AbstractSet<E>
      implements java.io.Serializable {
   private final CopyOnWriteArrayList<E> al;   //封装的CopyOnWriteArrayList
   public CopyOnWriteArraySet() {
      al = new CopyOnWriteArrayList<E>();
   }
   public boolean add(E e) {
      return al.addIfAbsent(e);    //不重复的，加进去
   }
   ...
}
```

5.4 ConcurrentLinkedQueue/Deque

前面详细分析了 AQS 内部的阻塞队列实现原理：基于双向链表，通过对 head/tail 进行 CAS 操作，实现入队和出队。

ConcurrentLinkedQueue 的实现原理和 AQS 内部的阻塞队列类似：同样是基于 CAS，同样是通过 head/tail 指针记录队列头部和尾部，但还是有稍许差别。

首先，它是一个单向链表，定义如下。

```
public class ConcurrentLinkedQueue<E> extends AbstractQueue<E>
    implements Queue<E>, java.io.Serializable {
private static class Node<E> {
volatile E item;
volatile Node<E> next;
...
}
private transient volatile Node<E> head;
private transient volatile Node<E> tail;
...
}
```

其次，在 AQS 的阻塞队列中，每次入队后，tail 一定后移一个位置；每次出队，head 一定前移一个位置，以保证 head 指向队列头部，tail 指向链表尾部。

但在 ConcurrentLinkedQueue 中，head/tail 的更新可能落后于节点的入队和出队，因为它不是直接对 head/tail 指针进行 CAS 操作的，而是对 Node 中的 item 进行操作。下面进行详细分析：

1. 初始化

初始的时候，如图 5-5 所示，head 和 tail 都执行一个 NULL 节点。对应的代码如下。

```
public ConcurrentLinkedQueue() {
    head = tail = new Node<E>(null);
}
```

图 5-5　初始状态

2. 入队列

代码如下所示。

```
public boolean offer(E e) {
checkNotNull(e);
final Node<E> newNode = new Node<E>(e);

for (Node<E> t = tail, p = t;;) {
    Node<E> q = p.next;
    if (q == null) {
        if (p.casNext(null, newNode)) {  //关键点：是对tail的next指针而不是
                                         //对tail指针执行CAS操作
            if (p != t)
                casTail(t, newNode);     //每入2个节点，后移一次tail指针，失败
                                         //也没有关系
            return true;
        }
    }
    else if (p == q)
        p = (t != (t = tail)) ? t : head;         //已经到达队列尾部
    else
        p = (p != t && t != (t = tail)) ? t : q;  //后移p指针
}
}
```

上面的入队其实是每次在队尾追加 2 个节点时，才移动一次 tail 节点，具体过程如图 5-6 所示。

初始的时候，队列中有 1 个节点 item1，tail 指向该节点，假设线程 1 要入队 item2 节点：

step1：p = tail，q = p.next = NULL。

step2：对 p 的 next 执行 CAS 操作，追加 item2，成功之后，p = tail。所以上面的 casTail 函数不会执行，直接返回。此时 tail 指针没有变化。

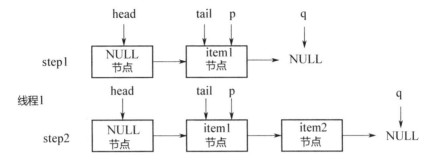

图 5-6　线程 1 入队 item2 节点

之后，假设线程 2 要入队 item3 节点，如图 5-7 所示。

step3：p = tail，q = p.next。

step4：q != NULL，因此不会入队新节点。p, q 都后移 1 位。

step5：q = NULL，对 p 的 next 执行 CAS 操作，入队 item3 节点。

step6：p != t，满足条件，执行上面的 casTail 操作，tail 后移 2 个位置，到达队列尾部。

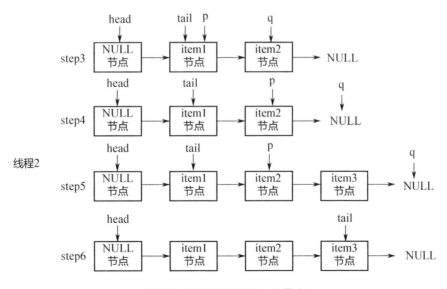

图 5-7　线程 2 入队 item3 节点

最后总结一下入队列的两个关键点：

（1）即使 tail 指针没有移动，只要对 p 的 next 指针成功进行 CAS 操作，就算成功入队列。

（2）只有当 p != tail 的时候，才会后移 tail 指针。也就是说，每连续追加 2 个节点，才后移 1 次 tail 指针。即使 CAS 失败也没关系，可以由下 1 个线程来移动 tail 指针。

3. 出队列

上面说了入队列之后，tail 指针不变化，那是否会出现入队列之后，要出队列却没有元素可出的情况呢？

```
public E poll() {
restartFromHead:
for (;;) {
    for (Node<E> h = head, p = h, q;;) {
        E item = p.item;

        if (item != null && p.casItem(item, null)) { //关键点：在出队列的时候，
                                                    //并没有移动 head 指针，而
                                                    //是把 item 置为了 NULL
            if (p != h) // 每产生 2 个 NULL 节点，才把 head 指针后移 2 位
                updateHead(h, ((q = p.next) != null) ? q : p);
            return item;
        }
        else if ((q = p.next) == null) {
            updateHead(h, p);
            return null;
        }
        else if (p == q)
            continue restartFromHead;
        else
            p = q;
    }
}
}
```

出队列的代码和入队列类似，也有 p、q 2 个指针，整个变化过程如图 5-8 所示。假设初始的时候 head 指向空节点，队列中有 item1、item2、item3 三个节点。

step1：p= head, q = p.next。p! =q。

step2：后移 p 指针，使得 p = q。

step3：出队列。关键点：此处并没有直接删除 item1 节点，只是把该节点的 item 通过 CAS

操作置为了 NULL。

step4：p != head，此时队列中有了 2 个 NULL 节点，再前移 1 次 head 指针，对其执行 updateHead 操作。

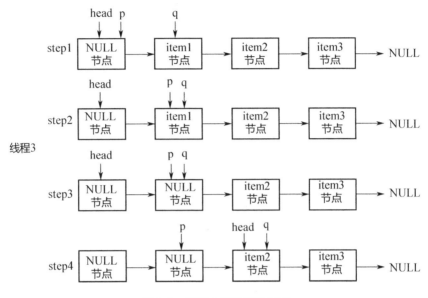

图 5-8　线程 3 出队列的过程

最后总结一下出队列的关键点：

（1）出队列的判断并非观察 tail 指针的位置，而是依赖于 head 指针后续的节点是否为 NULL 这一条件。

（2）只要对节点的 item 执行 CAS 操作，置为 NULL 成功，则出队列成功。即使 head 指针没有成功移动，也可以由下 1 个线程继续完成。

4. 队列判空

因为 head/tail 并不是精确地指向队列头部和尾部，所以不能简单地通过比较 head/tail 指针来判断队列是否为空，而是需要从 head 指针开始遍历，找第 1 个不为 NULL 的节点。如果找到，则队列不为空；如果找不到，则队列为空。代码如下所示。

```
public boolean isEmpty() {
return first() == null;      //寻找第 1 个不为 NULL 的节点
}
Node<E> first() {            //从 head 指针开始遍历，获取第 1 个不为 NULL 的节点
restartFromHead:
```

```
for (;;) {
    for (Node<E> h = head, p = h, q;;) {
        boolean hasItem = (p.item != null);
        if (hasItem || (q = p.next) == null) {
            updateHead(h, p);
            return hasItem ? p : null;
        }
        else if (p == q)
            continue restartFromHead;
        else
            p = q;
    }
}
```

5.5 ConcurrentHashMap

HashMap 通常的实现方式是"数组 + 链表",这种方式被称为"拉链法"。ConcurrentHashMap 在这个基本原理之上进行了各种优化,在 JDK 7 和 JDK 8 中的实现方式有很大差异,下面分开讨论。

5.5.1 JDK 7 中的实现方式

为了提高并发度,在 JDK7 中,一个 HashMap 被拆分为多个子 HashMap。每一个子 HashMap 称作一个 Segment,多个线程操作多个 Segment 相互独立,如图 5-9 所示。

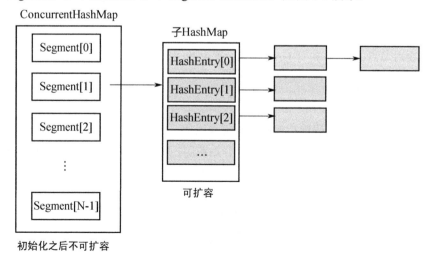

图 5-9 JDK 7 中 ConcurrentHashMap 数据结构示意图

具体来说，每个 Segment 都继承自 ReentrantLock，Segment 的数量等于锁的数量，这些锁彼此之间相互独立，即所谓的"分段锁"。代码如下所示。

```
public class ConcurrentHashMap<K, V> extends AbstractMap<K, V>
    implements ConcurrentMap<K, V>, Serializable {
...
static final class Segment<K,V> extends ReentrantLock implements Serializable {
...
}
final Segment<K,V>[] segments;   //ConcurrentHashMap 由多个子 HashMap (Segment)
                                 //组成
}
```

接下来由构造函数的分析入手，剖析 ConcurrentHashMap 的实现原理。

1. 构造函数分析

```
public ConcurrentHashMap(int initialCapacity, float loadFactor, int concurrencyLevel) {
    if (!(loadFactor > 0) || initialCapacity < 0 || concurrencyLevel <= 0)
        throw new IllegalArgumentException();
    if (concurrencyLevel > MAX_SEGMENTS)
        concurrencyLevel = MAX_SEGMENTS;

    int sshift = 0;
    int ssize = 1;
    while (ssize < concurrencyLevel) {     //保证并发度是 2 的整数次方
        ++sshift;
        ssize <<= 1;
    }
    this.segmentShift = 32 - sshift;
    this.segmentMask = ssize - 1;

    if (initialCapacity > MAXIMUM_CAPACITY)
        initialCapacity = MAXIMUM_CAPACITY;
    int c = initialCapacity / ssize; //初始总容量或数组个数，是每个 Segment 的初始大小
    if (c * ssize < initialCapacity)
        ++c;
    int cap = MIN_SEGMENT_TABLE_CAPACITY;
    while (cap < c)        //保证每个 Segment 的容量，也是 2 的整数次方
        cap <<= 1;
```

```
        Segment<K,V> s0 =
            new Segment<K,V>(loadFactor, (int)(cap * loadFactor),
                        (HashEntry<K,V>[])new HashEntry[cap]);   //构造第 0 个 Segment

        Segment<K,V>[] ss = (Segment<K,V>[])new Segment[ssize];  //关键的一句: Segment
                                                                 //数组的大小为 ssize,
                                                                 //也就是 2 的整数次方
        UNSAFE.putOrderedObject(ss, SBASE, s0); // 数组的第 0 个元素赋值为 s0
            this.segments = ss;
}
```

构造函数的第 3 个参数 concurrenyLevel，是"并发度"，也就是 Segment 数组的大小。这个值一旦在构造函数中设定，之后不能再扩容。为了提升 hash 的计算性能，会保证数组的大小始终是 2 的整数次方。例如设置 concurrentyLevel = 9，在构造函数里面会找到比 9 大且距 9 最近的 2 的整数次方，也就是 ssize = 16。对应 segmentShift、segmentMask 两个变量，是为了方便计算 hash 使用的。

初始的时候，如果不指定任何参数，就会使用默认值，代码如下所示。可以看到，默认的 Segment 数组大小是 16。

```
    public ConcurrentHashMap() {
        this(DEFAULT_INITIAL_CAPACITY, DEFAULT_LOAD_FACTOR, DEFAULT_CONCURRENCY_LEVEL);
}
        static final int DEFAULT_INITIAL_CAPACITY = 16;
static final float DEFAULT_LOAD_FACTOR = 0.75f;
static final int DEFAULT_CONCURRENCY_LEVEL = 16;
```

第 1 个参数，initialCapacity 是整个 ConcurrentHashMap 的初始大小。用 initialCapacity 除以 ssize，是每个 Segment 的初始大小。这里也会保证 Segment 里面 HashEntry[]数组的大小是 2 的整数次方。

第 2 个参数，loadFactor 即负载因子，传给了 Segment 内部。当每个 Segment 的元素个数达到一定阈值，进行 rehash。如图 5-9 所示，Segment 的个数不能扩容，但每个 Segment 的内部可以扩容。

2. put(..)函数分析

```
public V put(K key, V value) {
Segment<K,V> s;
if (value == null)
    throw new NullPointerException();
int hash = hash(key);    //把 key 映射到一个 32 位整数
```

```
        int j = (hash >>> segmentShift) & segmentMask;  //再把该整数映射到第 j 个 Segment
        if ((s = (Segment<K,V>)UNSAFE.getObject
                (segments, (j << SSHIFT) + SBASE)) == null)  //Segment[j] == null
            s = ensureSegment(j);                       //对第 j 个 Segment 进行初始化
        return s.put(key, hash, value, false);          //找到对应的 Segment[j]，调用其 put
}
```

由于 Segment 的个数是 2 的整数次方，假设其为 16，则 segmentShift = 32–ssshift = 32–4 = 28，segmentMask = ssize–1 = 16–1 = 15（参见上面的构造函数）。

hash 值是一个 32 位的整数，int j = (hash >>> segmentShift) & segmentMask 这行代码就是把 hash 值右移 28 位，再和 15 进行与操作，表达的意思是：以 hash 值的最高 4 位作为对应的 Segment 数组下标 j。

在上面的代码中没有加锁操作，因为锁是加在 s.put 内部的，也就是分段加锁。另外，多个线程可能同时调用 ensureSegment 对 Segment[j]进行初始化，在这个函数的内部要避免重复初始化，下面详细分析。

```
    private Segment<K,V> ensureSegment(int k) {
        final Segment<K,V>[] ss = this.segments;
        long u = (k << SSHIFT) + SBASE;  //下标 k 对应的内存地址偏移量为 u
        Segment<K,V> seg;
        if ((seg = (Segment<K,V>)UNSAFE.getObjectVolatile(ss, u)) == null) { //segments[k] = null
            Segment<K,V> proto = ss[0];  //以 ss[0]的参数为原型
            int cap = proto.table.length;
            float lf = proto.loadFactor;
            int threshold = (int)(cap * lf);
            HashEntry<K,V>[] tab = (HashEntry<K,V>[])new HashEntry[cap];
            if ((seg = (Segment<K,V>)UNSAFE.getObjectVolatile(ss, u))
                == null) {  // segments[k] = null
                Segment<K,V> s = new Segment<K,V>(lf, threshold, tab);
                while ((seg = (Segment<K,V>)UNSAFE.getObjectVolatile(ss, u))
                    == null) {
                    if (UNSAFE.compareAndSwapObject(ss, u, null, seg = s))
                        break;
                }
            }
        }
        return seg;
    }
```

上面这个函数的目的是当 segments[k] = null 时对其进行初始化。由于多个线程可能同时调

用该函数，UNSAFE.getObjectVolatile(ss, u) == null 出现了 3 次，在 3 个不同的时间点重复检查 segments[k] == null。但即使如此，也不能完全保证避免重复初始化，所以最后需要执行一个 CAS 操作 UNSAFE.compareAndSwapObject(ss, u, null, seg = s)保证只被初始化一次。保险起见，把这次 CAS 操作放在一个 while 循环里，保证出来的时候 segments[k]一定不为空。

segments[j]被成功地初始化了，下面进入内部，了解如何把元素放进去。

```
static final class Segment<K,V> extends ReentrantLock implements Serializable
{
    ...
    final V put(K key, int hash, V value, boolean onlyIfAbsent) {
        HashEntry<K,V> node = tryLock() ? null :
            scanAndLockForPut(key, hash, value);
        //当执行到这个地方时一定拿到锁了
        V oldValue;
        try {
            HashEntry<K,V>[] tab = table;
            int index = (tab.length - 1) & hash;   //因为 tab.length 为 2 的整数次方,这个
                                                   //地方等价于 hash 对 tab.length 取模
            HashEntry<K,V> first = entryAt(tab, index);   //定位到第 index 个 HashEntry
            for (HashEntry<K,V> e = first;;) {
                if (e != null) {
                    K k;
                    if ((k = e.key) == key ||
                        (e.hash == hash && key.equals(k))) {
                        oldValue = e.value;
                        if (!onlyIfAbsent) {
                            e.value = value;
                            ++modCount;           //修改次数累加
                        }
                        break;       //key 相等或者 hash 值相等,不会重复插入,直接返回
                    }
                    e = e.next;    //遍历链表
                }
                else {       //已经遍历到链表尾部,没有发现重复元素
                    if (node != null)    //在上面的 scanAndLockForPut 里面,已经建好了节点
                        node.setNext(first);    //把 node 插入链表头部
                    else
                        node = new HashEntry<K,V>(hash, key, value, first);   //新建 node
                                                                               //插入链表头部
                    int c = count + 1;
                    if (c > threshold && tab.length < MAXIMUM_CAPACITY)
```

```
                    rehash(node);          //超出阈值，扩容
                else
                    setEntryAt(tab, index, node);   //把 node 赋值给 tab[index]
                ++modCount;
                count = c;
                oldValue = null;
                break;
            }
        }
    } finally {
        unlock();
    }
    return oldValue;
}
```

关于上面的代码，有几点说明：

（1）在 Segment 里面有 2 个 count 变量，count 与 modCount，前者表示元素的个数，后者表示修改的次数。当待 put 的元素、key 值或者 hash 值和链表中某个节点相等时，不会重复插入新节点。此时再进一步判断，如果 onlyIfAbsent = true，则什么都不做；如果 onlyIfAbsent = false，则修改该节点的 value，同时 modCount 累加一次。

（2）进入上面的 else 分支，说明已经遍历到了链表尾部，并且没有发现与 key 值或者 hash 值相等的节点，此时在链表头部插入 1 个新节点，并把 table[index]赋值为该节点。因为 first 就是链表头部，所以直接把 node 的 next 指针指向 first 就可以了。

（3）在函数的开始，加锁的时候，进行了一次优化。

```
HashEntry<K,V> node = tryLock() ? null : scanAndLockForPut(key, hash, value);
```

如果 tryLock()成功，即拿到锁，则进入下面的代码；如果 tryLock()不成功，也不能闲着，那进入 scanAndLockForPut(key, hash, value)做什么呢？

```
private HashEntry<K,V> scanAndLockForPut(K key, int hash, V value) {
    HashEntry<K,V> first = entryForHash(this, hash);
    HashEntry<K,V> e = first;
    HashEntry<K,V> node = null;
    int retries = -1;
    while (!tryLock()) {
        HashEntry<K,V> f;
        if (retries < 0) {
            if (e == null) {
                if (node == null)   //创建一个新节点
                    node = new HashEntry<K,V>(hash, key, value, null);
```

```
                retries = 0;
            }
            else if (key.equals(e.key))
                retries = 0;
            else
                e = e.next;
        }
        else if (++retries > MAX_SCAN_RETRIES) {    //自旋
            lock();    //阻塞
            break;
        }
        else if ((retries & 1) == 0 &&
                (f = entryForHash(this, hash)) != first) {
            e = first = f;
            retries = -1;
        }
    }
    return node;
}
```

上面的函数看似复杂，实则只是做了两件事：一是拿不到锁，不立即阻塞，而是先自旋，若自旋到一定次数仍未拿到锁，再调用lock()阻塞；二是在自旋的过程中遍历了链表，若发现没有重复的节点，则提前新建一个节点，为后面再插入节省时间。

3. 扩容

在上面的代码中提到了，超过一定的阈值后，Segment 内部会进行扩容，代码如下。

```
private void rehash(HashEntry<K,V> node) {
    HashEntry<K,V>[] oldTable = table;
    int oldCapacity = oldTable.length;
    int newCapacity = oldCapacity << 1;    //在旧容量的基础上扩容二倍
    threshold = (int)(newCapacity * loadFactor);
    HashEntry<K,V>[] newTable =
        (HashEntry<K,V>[]) new HashEntry[newCapacity];
    int sizeMask = newCapacity - 1;

    for (int i = 0; i < oldCapacity ; i++) {
        HashEntry<K,V> e = oldTable[i];
        if (e != null) {
            HashEntry<K,V> next = e.next;
            int idx = e.hash & sizeMask;    //如果一个节点之前在第i个位置,那么在新的hash
                                            //表中,一定处于i或者i+oldCapacity位置
            if (next == null)
```

```
                    newTable[idx] = e;
                else {
                    HashEntry<K,V> lastRun = e;
                    int lastIdx = idx;
                    for (HashEntry<K,V> last = next;
                         last != null;
                         last = last.next) {
                        int k = last.hash & sizeMask;
                        if (k != lastIdx) {
                            lastIdx = k;   //寻找链表中最后一个hash值不等于lastIdx的元素
                            lastRun = last;
                        }
                    }
                    newTable[lastIdx] = lastRun;   //关键的一行：把在lastRun之后的链表元
                                                   //素直接链接到新hash表中的lastIdx位置
                                                   //在lastRun之前的所有链表元素，需要在
                                                   //新的位置逐个拷贝
                    for (HashEntry<K,V> p = e; p != lastRun; p = p.next) {
                        V v = p.value;
                        int h = p.hash;
                        int k = h & sizeMask;
                        HashEntry<K,V> n = newTable[k];
                        newTable[k] = new HashEntry<K,V>(h, p.key, v, n);
                    }
                }
            }
        }
    int nodeIndex = node.hash & sizeMask;    //把新节点加入新的Hash表
        node.setNext(newTable[nodeIndex]);
    newTable[nodeIndex] = node;
    table = newTable;
}
```

关于该扩容函数，有几点需要说明：

（1）函数的参数，也就是将要加入的最新节点。在扩容完成之后，把该节点加入新的 Hash 表。

（2）整个数组的长度是 2 的整数次方，每次按二倍扩容，而 hash 函数就是对数组长度取模，即 node.hash & sizeMask。因此，如果元素之前处于第 i 个位置，当再次 hash 时，必然处于第 i 个或者第 i + oldCapacity 个位置。

（3）上面的扩容进行了一次优化，并没有对元素依次拷贝，而是先找到 lastRun 位置，也就是 for 循环。lastRun 到链表末尾的所有元素，其 hash 值没有改变，所以不需要依次重新拷贝，只需把这部分链表链接到新链表所对应的位置就可以，也就是 newTable[lastIdx] = lastRun。lastRun 之前的元素则需要依次拷贝。

因此，即使把一段 for 循环去掉，整个程序的逻辑仍然是正确的。

4. get 实现分析

```
public V get(Object key) {
Segment<K,V> s;
HashEntry<K,V>[] tab;
int h = hash(key);
long u = (((h >>> segmentShift) & segmentMask) << SSHIFT) + SBASE;
//第 1 次 hash
if ((s = (Segment<K,V>)UNSAFE.getObjectVolatile(segments, u)) != null &&
    (tab = s.table) != null) {
    for (HashEntry<K,V> e = (HashEntry<K,V>) UNSAFE.getObjectVolatile
            (tab, ((long)(((tab.length - 1) & h)) << TSHIFT) + TBASE);
            //第 2 次 hash
         e != null; e = e.next) {
        K k;
        if ((k = e.key) == key || (e.hash == h && key.equals(k)))
            return e.value;
    }
}
return null;
}
```

整个 get 过程也就是两次 hash：

第一次 hash，函数为(h >>>segmentShift) & segmentMask，计算出所在的 Segment；

第二次 hash，函数为 h & (tab.length -1)，即 h 对数组长度取模，找到 Segment 里面对应的 HashEntry 数组下标，然后遍历该位置的链表。

整个读的过程完全没有加锁，而是使用了 UNSAFE.getObjectVolatile 函数。

5.5.2　JDK 8 中的实现方式

JDK 8 的实现有很大变化，首先是没有了分段锁，所有数据都放在一个大的 HashMap 中；其次是引入了红黑树，其原理如图 5-10 所示。

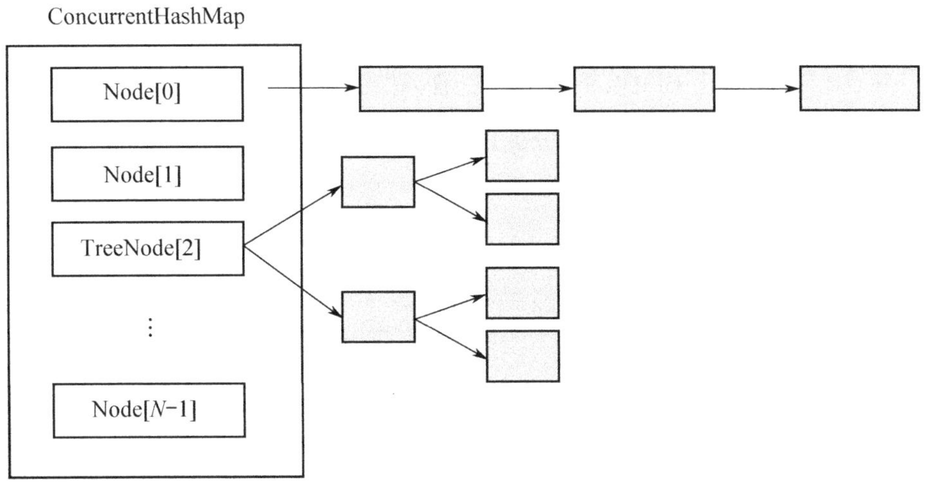

图 5-10 JDK 8 中 ConcurrentHashMap 实现原理示意图

如果头节点是 Node 类型，则尾随它的就是一个普通的链表；如果头节点是 TreeNode 类型，它的后面就是一棵红黑树，TreeNode 是 Node 的子类。

链表和红黑树之间可以相互转换：初始的时候是链表，当链表中的元素超过某个阈值时，把链表转换成红黑树；反之，当红黑树中的元素个数小于某个阈值时，再转换为链表。

那为什么 JDK 8 要做这种改变呢？在 JDK 7 中的分段锁，有三个好处：

（1）减少 Hash 冲突，避免一个槽里有太多元素。

（2）提高读和写的并发度。段与段之间相互独立。

（3）提供扩容的并发度。扩容的时候，不是整个 ConcurrentHashMap 一起扩容，而是每个 Segment 独立扩容。

针对这三个好处，我们来看一下在 JDK 8 中相应的处理方式：

（1）使用红黑树，当一个槽里有很多元素时，其查询和更新速度会比链表快很多，Hash 冲突的问题由此得到较好的解决。

（2）加锁的粒度，并非整个 ConcurrentHashMap，而是对每个头节点分别加锁，即并发度，就是 Node 数组的长度，初始长度为 16，和在 JDK 7 中初始 Segment 的个数相同。

（3）并发扩容，这是难度最大的。在 JDK 7 中，一旦 Segment 的个数在初始化的时候确立，不能再更改，并发度被固定。之后只是在每个 Segment 内部扩容，这意味着每个 Segment 独立扩容，互不影响，不存在并发扩容的问题。但在 JDK 8 中，相当于只有 1 个 Segment，当一个线程要扩容 Node 数组的时候，其他线程还要读写，因此处理过程很复杂，后面会详细分析。

由上述对比可以总结出来：JDK 8 的实现一方面降低了 Hash 冲突，另一方面也提升了并发度。

下面从构造函数开始，一步步深入分析其实现过程。

1. 构造函数分析

```
public ConcurrentHashMap(int initialCapacity) {
if (initialCapacity < 0)
    throw new IllegalArgumentException();
int cap = ((initialCapacity >= (MAXIMUM_CAPACITY >>> 1)) ?
        MAXIMUM_CAPACITY :
        tableSizeFor(initialCapacity + (initialCapacity >>> 1) + 1));
this.sizeCtl = cap;
}
```

在上面的代码中，变量 cap 就是 Node 数组的长度，保持为 2 的整数次方。tableSizeFor(..) 函数是根据传入的初始容量，计算出一个合适的数组长度。具体而言：1.5 倍的初始容量 + 1，再往上取最接近的 2 的整数次方，作为数组长度 cap 的初始值。

这里的 sizeCtl，其含义是用于控制在初始化或者并发扩容时候的线程数，只不过其初始值设置成 cap。

2. 初始化

在上面的构造函数里只计算了数组的初始大小，并没有对数组进行初始化。当多个线程都往里面放入元素的时候，再进行初始化。这就存在一个问题：多个线程重复初始化。下面看一下是如何处理的。

```
private final Node<K,V>[] initTable() {
Node<K,V>[] tab; int sc;
while ((tab = table) == null || tab.length == 0) {
    if ((sc = sizeCtl) < 0)
        Thread.yield(); //sizeCtl = -1，自旋等待
    else if (U.compareAndSwapInt(this, SIZECTL, sc, -1)) {   //关键:把sizeCtl
                                                              //设为-1
        try {
            if ((tab = table) == null || tab.length == 0) {
                int n = (sc > 0) ? sc : DEFAULT_CAPACITY;
                Node<K,V>[] nt = (Node<K,V>[])new Node<?,?>[n];   //初始化
                table = tab = nt;
                sc = n - (n >>> 2);   //sizeCtl 并非表示数组长度，所以初始化成功之后，
                                      //就不再等于数组长度，而是 n-(n >>> 2) =
                                      //n-n/4 = 0.75n，表示下一次扩容的阈值
```

```
            }
        } finally {
            sizeCtl = sc;               //把 sizeCtl 再设回去
        }
            break;
        }
    }
    return tab;
}
```

通过上面的代码可以看到，多个线程的竞争是通过对 sizeCtl 进行 CAS 操作实现的。如果某个线程成功地把 sizeCtl 设置为-1，它就拥有了初始化的权利，进入初始化的代码模块，等到初始化完成，再把 sizeCtl 设置回去；其他线程则一直执行 while 循环，自旋等待，直到数组不为 null，即当初始化结束时，退出整个函数。

因为初始化的工作量很小，所以此处选择的策略是让其他线程一直等待，而没有帮助其初始化。

3. put(..)实现分析

```
public V put(K key, V value) {
    return putVal(key, value, false);
}
final V putVal(K key, V value, boolean onlyIfAbsent) {
    if (key == null || value == null) throw new NullPointerException();
    int hash = spread(key.hashCode());
    int binCount = 0;
    for (Node<K,V>[] tab = table;;) {
        Node<K,V> f; int n, i, fh;
        if (tab == null || (n = tab.length) == 0)
            tab = initTable();      //分支 1：整个数组初始化
        else if ((f = tabAt(tab, i = (n - 1) & hash)) == null) {  //分支 2：第[i]
                                                                  //个元素初始化
            if (casTabAt(tab, i, null,
                    new Node<K,V>(hash, key, value, null)))
                break;
        }
        else if ((fh = f.hash) == MOVED)
            tab = helpTransfer(tab, f);     //分支 3：扩容
        else {                              //分支 4：放入元素
            V oldVal = null;
            synchronized (f) {              //关键的一句：加锁
                if (tabAt(tab, i) == f) {
```

```java
            if (fh >= 0) {   //链表
                binCount = 1;
                for (Node<K,V> e = f;; ++binCount) {
                    K ek;
                    if (e.hash == hash &&
                        ((ek = e.key) == key ||
                         (ek != null && key.equals(ek)))) {
                        oldVal = e.val;
                        if (!onlyIfAbsent)
                            e.val = value;
                        break;
                    }
                    Node<K,V> pred = e;
                    if ((e = e.next) == null) {
                        pred.next = new Node<K,V>(hash, key,
                                                  value, null);
                        break;
                    }
                }
            }
            else if (f instanceof TreeBin) {   //红黑树
                Node<K,V> p;
                binCount = 2;
                if ((p = ((TreeBin<K,V>)f).putTreeVal(hash, key,
                                               value)) != null) {
                    oldVal = p.val;
                    if (!onlyIfAbsent)
                        p.val = value;
                }
            }
        }
    }
    if (binCount != 0) {    //如果是链表，则上面的binCount会从1一直累加
        if (binCount >= TREEIFY_THRESHOLD)
            treeifyBin(tab, i);   //超出阈值，转换为红黑树
        if (oldVal != null)
            return oldVal;
        break;
    }
}
}
```

```
        addCount(1L, binCount);    //总元素个数累加 1
        return null;
}
```

上面的 for 循环有 4 个大的分支：

第 1 个分支，是整个数组的初始化，前面已讲；

第 2 个分支，是所在的槽为空，说明该元素是该槽的第一个元素，直接新建一个头节点，然后返回；

第 3 个分支，说明该槽正在进行扩容，帮助其扩容；

第 4 个分支，就是把元素放入槽内。槽内可能是一个链表，也可能是一棵红黑树，通过头节点的类型可以判断是哪一种。第 4 个分支是包裹在 synchronized (f)里面的，f 对应的数组下标位置的头节点，意味着每个数组元素有一把锁，并发度等于数组的长度。

上面的 binCount 表示链表的元素个数，当这个数目超过 TREEIFY_THRESHOLD = 8 时，把链表转换成红黑树，也就是 treeifyBin(tab, i)函数。但在这个函数内部，不一定需要进行红黑树转换，可能只做扩容操作，所以接下来从扩容讲起。

4. 扩容

扩容的实现是最复杂的，下面从 treeifyBin(tab, i)讲起。

```
private final void treeifyBin(Node<K,V>[] tab, int index) {
Node<K,V> b; int n, sc;
if (tab != null) {
    if ((n = tab.length) < MIN_TREEIFY_CAPACITY)
        tryPresize(n << 1);    //数组长度小于阈值 64，不做红黑树转换，直接扩容
    else if ((b = tabAt(tab, index)) != null && b.hash >= 0) {
        synchronized (b) {    //链表转换成红黑树
            if (tabAt(tab, index) == b) {
                TreeNode<K,V> hd = null, tl = null;
                for (Node<K,V> e = b; e != null; e = e.next) {
                    TreeNode<K,V> p =        //遍历链表，构建红黑树
                        new TreeNode<K,V>(e.hash, e.key, e.val,
                                         null, null);
                    if ((p.prev = tl) == null)
                        hd = p;
                    else
                        tl.next = p;
                    tl = p;
                }
```

```
                    setTabAt(tab, index, new TreeBin<K,V>(hd));
                }
            }
        }
    }
}
```

在上面的代码中,MIN_TREEIFY_CAPACITY = 64,意味着当数组的长度没有超过 64 的时候,数组的每个节点里都是链表,只会扩容,不会转换成红黑树。只有当数组长度大于或等于 64 时,才考虑把链表转换成红黑树。

在 tryPresize(int size) 内部调用了一个核心函数 transfer(Node<K,V>[] tab, Node<K,V>[] nextTab),先从这个函数的分析说起。

```
private final void transfer(Node<K,V>[] tab, Node<K,V>[] nextTab) {
int n = tab.length, stride;

if ((stride = (NCPU > 1) ? (n >>> 3) / NCPU : n) < MIN_TRANSFER_STRIDE)
    stride = MIN_TRANSFER_STRIDE;        //计算步长
if (nextTab == null) {                    //初始化新 HashMap
    try {
        @SuppressWarnings("unchecked")
        Node<K,V>[] nt = (Node<K,V>[])new Node<?,?>[n << 1];  //扩容 2 倍
        nextTab = nt;
    } catch (Throwable ex) {
        sizeCtl = Integer.MAX_VALUE;
        return;
    }
    nextTable = nextTab;
    transferIndex = n;    //初始的 transferIndex 为旧 HashMap 的数组长度
}

int nextn = nextTab.length;
ForwardingNode<K,V> fwd = new ForwardingNode<K,V>(nextTab);
boolean advance = true;
boolean finishing = false;
//这里的 i 为遍历的下标,bound 为遍历的边界。如果成功拿到一个任务,则 i = nextIndex -1,
//bound = nextIndex -stride; 如果拿不到任务,则 i = 0, bound = 0
for (int i = 0, bound = 0;;) {
    Node<K,V> f; int fh;
//advance 表示在从 i = transferIndex-1 遍历到 bound 位置的过程中,是否一直继续
    while (advance) {        //这个地方较难理解: 3 个分支里面都是 advance = false,意味着
                             //若 3 个分支都不执行,才可能一直执行 while 循环。目的在于,当
```

```java
                              //对 transferIndex 执行 CAS 操作不成功的时候,需要自旋,以期拿
                              //到一个 stride 的迁移任务。
            int nextIndex, nextBound;
            if (--i >= bound || finishing)    //对数组的遍历,通过这里的- - i 进行。如果
                                              //成功执行了 - - i,就不用继续 while 循环
                                              //了。因为每次 advance 只能进一步
                advance = false;
            else if ((nextIndex = transferIndex) <= 0) {   //transferIndex < = 0,
                                                           //整个 HashMap 完成
                i = -1;
                advance = false;
            }
            else if (U.compareAndSwapInt     //对 transferIndex 进行 CAS 操作,也就是
                                             //为当前线程分配 1 个 stride。CAS 操作成功,
                                             //线程成功拿到一个 stride 的迁移任务; CAS
                                             //操作不成功,线程没抢到任务,会继续执行
                                             //while 循环,自旋
                    (this, TRANSFERINDEX, nextIndex,
                     nextBound = (nextIndex > stride ?
                                  nextIndex - stride : 0))) {
                bound = nextBound;
                i = nextIndex - 1;
                advance = false;
            }
        }
//i 已经越界,整个 HashMap 已经遍历完成
        if (i < 0 || i >= n || i + n >= nextn) {
            int sc;
            if (finishing) {    //finishing 表示整个 HashMap 扩容完成
                nextTable = null;
                table = nextTab;   //把 nextTab 赋值给当前 table
                sizeCtl = (n << 1) - (n >>> 1);
                return;
            }
            if (U.compareAndSwapInt(this, SIZECTL, sc = sizeCtl, sc - 1)) {
                if ((sc - 2) != resizeStamp(n) << RESIZE_STAMP_SHIFT)
                    return;
                finishing = advance = true;
                i = n;
            }
        }
        else if ((f = tabAt(tab, i)) == null)   //tab[i]迁移完毕,赋值一个 ForwardingNode
```

```
                advance = casTabAt(tab, i, null, fwd);
            else if ((fh = f.hash) == MOVED)   //tab[i]的位置已经在迁移过程中
                advance = true;
            else {   //对tab[i]进行迁移操作，tab[i]可能是一个链表或者红黑树
                synchronized (f) {
                    if (tabAt(tab, i) == f) {
                        Node<K,V> ln, hn;
                        if (fh >= 0) {   //链表
                            int runBit = fh & n;
                            Node<K,V> lastRun = f;
                            for (Node<K,V> p = f.next; p != null; p = p.next) {
                                int b = p.hash & n;
                                if (b != runBit) {
                                    runBit = b;
                                    lastRun = p;   //意味着在lastRun之后的所有元素，hash值
                                                   //都是一样的，记录下这个最后的位置
                                }
                            }
                            if (runBit == 0) {
                                ln = lastRun;   //类似于JDK 7的链表迁移的优化做法
                                hn = null;
                            }
                            else {
                                hn = lastRun;
                                ln = null;
                            }
                            for (Node<K,V> p = f; p != lastRun; p = p.next) {
                                int ph = p.hash; K pk = p.key; V pv = p.val;
                                if ((ph & n) == 0)
                                    ln = new Node<K,V>(ph, pk, pv, ln);
                                else
                                    hn = new Node<K,V>(ph, pk, pv, hn);
                            }
                            setTabAt(nextTab, i, ln);
                            setTabAt(nextTab, i + n, hn);
                            setTabAt(tab, i, fwd);
                            advance = true;
                        }
                        else if (f instanceof TreeBin) {   //红黑树，迁移办法和链表类似
                            TreeBin<K,V> t = (TreeBin<K,V>)f;
                            TreeNode<K,V> lo = null, loTail = null;
                            TreeNode<K,V> hi = null, hiTail = null;
```

```
            int lc = 0, hc = 0;
            for (Node<K,V> e = t.first; e != null; e = e.next) {
                int h = e.hash;
                TreeNode<K,V> p = new TreeNode<K,V>
                    (h, e.key, e.val, null, null);
                if ((h & n) == 0) {
                    if ((p.prev = loTail) == null)
                        lo = p;
                    else
                        loTail.next = p;
                    loTail = p;
                    ++lc;
                }
                else {
                    if ((p.prev = hiTail) == null)
                        hi = p;
                    else
                        hiTail.next = p;
                    hiTail = p;
                    ++hc;
                }
            }
            ln = (lc <= UNTREEIFY_THRESHOLD) ? untreeify(lo) :
                (hc != 0) ? new TreeBin<K,V>(lo) : t;
            hn = (hc <= UNTREEIFY_THRESHOLD) ? untreeify(hi) :
                (lc != 0) ? new TreeBin<K,V>(hi) : t;
            setTabAt(nextTab, i, ln);
            setTabAt(nextTab, i + n, hn);
            setTabAt(tab, i, fwd);
            advance = true;
        }
    }
}
```

该函数非常复杂，下面一步步分析：

（1）扩容的基本原理如图 5-11 所示，首先建一个新的 HashMap，其数组长度是旧数组长度的 2 倍，然后把旧的元素逐个迁移过来。所以，上面的函数参数有 2 个，第 1 个参数 tab 是扩容之前的 HashMap，第 2 个参数 nextTab 是扩容之后的 HashMap。当 nextTab = null 的时候，函数最初会对nextTab进行初始化。这里有一个关键点要说明：该函数会被多个线程调用，所以每个线程只是扩容旧的HashMap部分，这就涉及如何划分任务的问题。

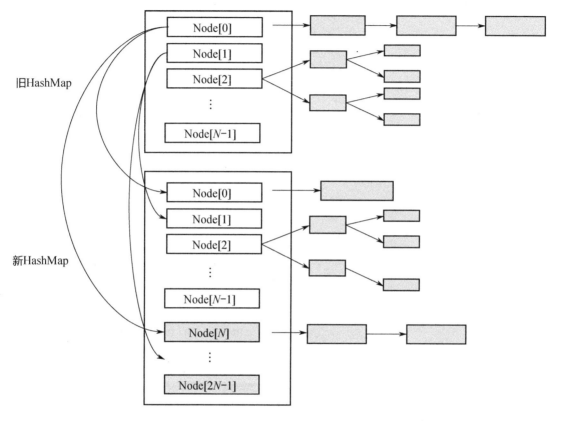

图 5-11 扩容的基本原理

（2）图 5-12 所示为多个线程并行扩容-任务划分示意图。旧数组的长度是 N，每个线程扩容一段，一段的长度用变量 stride（步长）来表示，transferIndex 表示了整个数组扩容的进度。

stride 的计算公式如上面的代码所示，即：在单核模式下直接等于 n，因为在单核模式下没有办法多个线程并行扩容，只需要 1 个线程来扩容整个数组；人在多核模式下为 (n>>>3)/NCPU，并且保证步长的最小值是 16。显然，需要的线程个数约为 n/stride。

```
if ((stride = (NCPU > 1) ? (n >>> 3) / NCPU : n) < MIN_TRANSFER_STRIDE)
    stride = MIN_TRANSFER_STRIDE;
```

transferIndex 是 ConcurrentHashMap 的一个成员变量，记录了扩容的进度。初始值为 n，从大到小扩容，每次减 stride 个位置，最终减至 n<= 0，表示整个扩容完成。因此，从[0, transferIndex-1]的位置表示还没有分配到线程扩容的部分，从[transfexIndex,n-1]的位置表示已经分配给某个线程

进行扩容，当前正在扩容中，或者已经扩容成功。

因为 transferIndex 会被多个线程并发修改，每次减 stride，所以需要通过 CAS 进行操作，如下面的代码所示。

```
else if (U.compareAndSwapInt
            (this, TRANSFERINDEX, nextIndex,
             nextBound = (nextIndex > stride ?
                          nextIndex - stride : 0)))
```

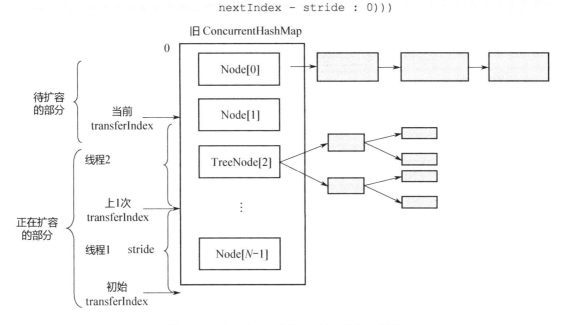

图 5-12 多个线程并行扩容–任务划分示意图

（3）在扩容未完成之前，有的数组下标对应的槽已经迁移到了新的 HashMap 里面，有的还在旧的 HashMap 里面。这个时候，所有调用 get(k,v) 的线程还是会访问旧 HashMap，怎么处理呢？图 5-13 所示为扩容过程中的转发示意图：当 Node[0] 已经迁移成功，而其他 Node 还在迁移过程中时，如果有线程要读取 Node[0] 的数据，就会访问失败。为此，新建一个 ForwardingNode，即转发节点，在这个节点里面记录的是新的 ConcurrentHashMap 的引用。这样，当线程访问到 ForwardingNode 之后，会去查询新的 ConcurrentHashMap。

（4）因为数组的长度 tab.length 是 2 的整数次方，每次扩容又是 2 倍。而 Hash 函数是 hashCode % tab.length，等价于 hashCode &(tab.length–1)。这意味着：处于第 i 个位置的元素，在新的 Hash 表的数组中一定处于第 i 个或者第 i+n 个位置，如图 5-11 所示。举个简单的例子：假设数组长度是 8，扩容之后是 16：

若 hashCode = 5，5 % 8 = 0，扩容后，5 % 16 = 0，位置保持不变；

若 hashCode = 24，24 % 8 = 0，扩容后，24 % 16 = 8，后移 8 个位置；

若 hashCode = 25，25 % 8 = 1，扩容后，25 % 16 = 9，后移 8 个位置；

若 hashCode = 39，39 % 8 = 7，扩容后，39 % 8 = 7，位置保持不变；

……

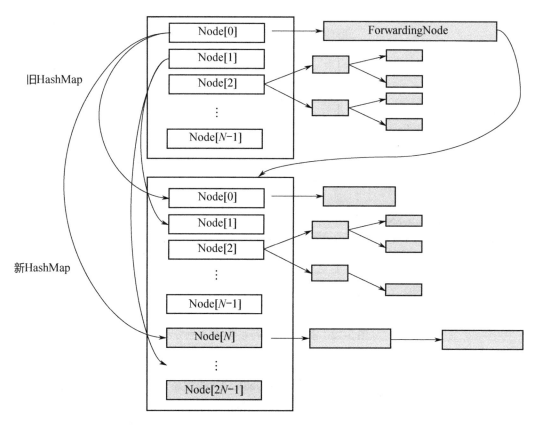

图 5-13　扩容过程中的转发示意图

正因为有这样的规律，所以如下有代码：

```
setTabAt(nextTab, i, ln);
setTabAt(nextTab, i + n, hn);
setTabAt(tab, i, fwd);
```

也就是把 tab[i] 位置的链表或红黑树重新组装成两部分，一部分链接到 nextTab[i] 的位置，一部分链接到 nextTab[i+n] 的位置，如图 5-11 所示。然后把 tab[i] 的位置指向一个 ForwardingNode

节点。

同时，当 tab[i] 后面是链表时，使用类似于 JDK 7 中在扩容时的优化方法，从 lastRun 往后的所有节点，不需依次拷贝，而是直接链接到新的链表头部。从 lastRun 往前的所有节点，需要依次拷贝。

了解了核心的迁移函数 transfer(tab, nextTab)，再回头看 tryPresize(int size) 函数。这个函数的输入是整个 Hash 表的元素个数，在函数里面，根据需要对整个 Hash 表进行扩容。想要看明白这个函数，需要透彻地理解 sizeCtl 变量，下面这段注释摘自源码。

```
/**
 * Table initialization and resizing control.  When negative, the
 * table is being initialized or resized: -1 for initialization,
 * else -(1 + the number of active resizing threads).  Otherwise,
 * when table is null, holds the initial table size to use upon
 * creation, or 0 for default. After initialization, holds the
 * next element count value upon which to resize the table.
 */
private transient volatile int sizeCtl;
```

当 sizeCtl = −1 时，表示整个 HashMap 正在初始化；

当 sizeCtl = 某个其他负数时，表示多个线程在对 HashMap 做并发扩容；

当 sizeCtl = cap 时，tab = null，表示未初始之前的初始容量（如上面的构造函数所示）；

扩容成功之后，sizeCtl 存储的是下一次要扩容的阈值，即上面初始化代码中的 n−(n >>> 2) = 0.75n。

所以，sizeCtl 变量在 Hash 表处于不同状态时，表达不同的含义。明白了这个道理，再来看上面的 tryPresize(int size) 函数。

```
private final void tryPresize(int size) {
    int c = (size >= (MAXIMUM_CAPACITY >>> 1)) ? MAXIMUM_CAPACITY :
        tableSizeFor(size + (size >>> 1) + 1);   //根据元素个数计算数组大小
    int sc;
    while ((sc = sizeCtl) >= 0) {
        Node<K,V>[] tab = table; int n;
        if (tab == null || (n = tab.length) == 0) {   //Hash 表初始化，和上面初始化
                                                      //的时候一样
            n = (sc > c) ? sc : c;
            if (U.compareAndSwapInt(this, SIZECTL, sc, -1)) {
                try {
                    if (table == tab) {
```

```
                @SuppressWarnings("unchecked")
                Node<K,V>[] nt = (Node<K,V>[])new Node<?,?>[n];
                table = nt;
                sc = n - (n >>> 2);    //即n-n/4 = 0.75n,下一次扩容的阈值
            }
        } finally {
            sizeCtl = sc;
        }
    }
    else if (c <= sc || n >= MAXIMUM_CAPACITY)
        break;
    else if (tab == table) {
        int rs = resizeStamp(n);
        if (sc < 0) {   //sc < 0,说明多个线程正在进行并发扩容
            Node<K,V>[] nt;
            if ((sc >>> RESIZE_STAMP_SHIFT) != rs || sc == rs + 1 ||
                sc == rs + MAX_RESIZERS || (nt = nextTable) == null ||
                transferIndex <= 0)      //扩容结束
                break;
            if (U.compareAndSwapInt(this, SIZECTL, sc, sc + 1))
                transfer(tab, nt);       //帮着扩容
        }
        else if (U.compareAndSwapInt(this, SIZECTL, sc,
                                     (rs << RESIZE_STAMP_SHIFT) + 2))
            transfer(tab, null);         //第一次扩容
    }
}
```

tryPresize(int size)是根据期望的元素个数对整个Hash表进行扩容,核心是调用transfer函数。在第一次扩容的时候,sizeCtl会被设置成一个很大的负数U.compareAndSwapInt(this, SIZECTL, sc,(rs << RESIZE_STAMP_SHIFT) + 2);之后每一个线程扩容的时候,sizeCtl 就加 1,U.compareAndSwapInt(this, SIZECTL, sc, sc + 1),待扩容完成之后,sizeCtl 减 1。

5.6 ConcurrentSkipListMap/Set

ConcurrentHashMap 是一种 key 无序的 HashMap,HashMap 则是 key 有序的,实现了NavigableMap 接口,此接口又继承了 SortedMap 接口。

5.6.1 ConcurrentSkipListMap

1. 为什么要使用 SkipList 实现 Map？

在 Java 的 util 包中，有一个非线程安全的 HashMap，也就是 TreeMap，是 key 有序的，基于红黑树实现。

而在 Concurrent 包中，提供的 key 有序的 HashMap，也就是 ConcurrentSkipListMap，是基于 SkipList（跳查表）来实现的。这里为什么不用红黑树，而用跳查表来实现呢？

借用 Doug Lea 的原话：

The reason is that there are no known efficient lock-free insertion and deletion algorithms for search trees.

也就是目前计算机领域还未找到一种高效的、作用在树上的、无锁的、增加和删除节点的办法。

那为什么 SkipList 可以无锁地实现节点的增加、删除呢？这要从无锁链表的实现说起。

2. 无锁链表

Doug Lea 在注释中引用了一篇无锁链表的论文：*A pragmatic implementation of non-blocking linked lists*。

表面上看，无锁链表是很简单的，根本不需要写一篇论文来专门论述。在上文中讲解 AQS 时，曾反复用到无锁队列，其实现也是链表。究竟二者的区别在哪呢？

上文所讲的无锁队列、栈，都是只在队头、队尾进行 CAS 操作，通常不会有问题。如果在链表的中间进行插入或删除操作，按照通常的 CAS 做法，就会出现问题！

关于这个问题，Doug Lea 的论文中有清晰的论述，此处引用如下：

操作 1：在节点 10 后面插入节点 20。如图 5-14 所示，首先把节点 20 的 next 指针指向节点 30，然后对节点 10 的 next 指针执行 CAS 操作，使其指向节点 20 即可。

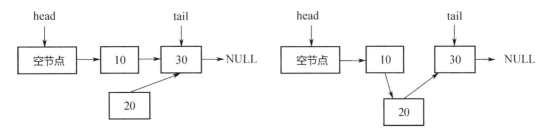

图 5-14 在节点 10 和节点 30 之间插入节点 20

操作 2：删除节点 10。如图 5-15 所示，只需把头节点的 next 指针，进行 CAS 操作到节点 30 即可。

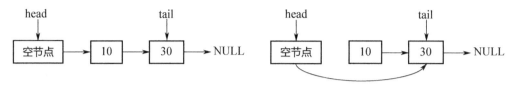

图 5-15　删除节点 10

但是，如果两个线程同时操作，一个删除节点 10，一个要在节点 10 后面插入节点 20。并且这两个操作都各自是 CAS 的，此时就会出现问题。如图 5-16 所示，删除节点 10，会同时把新插入的节点 20 也删除掉！这个问题超出了 CAS 的解决范围。

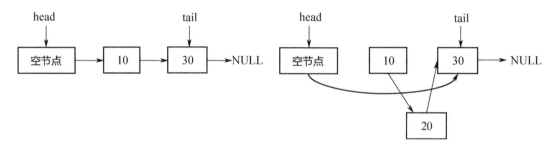

图 5-16　两个线程同时操作

为什么会出现这个问题呢？

究其原因：在删除节点 10 的时候，实际受到操作的是节点 10 的前驱，也就是头节点。节点 10 本身没有任何变化。故而，再往节点 10 后插入节点 20 的线程，并不知道节点 10 已经被删除了！

针对这个问题，在论文中提出了如下的解决办法，如图 5-17 所示，把节点 10 的删除分为两 2 步：

第一步，把节点 10 的 next 指针，mark 成删除，即软删除；

第二步，找机会，物理删除。

做标记之后，当线程再往节点 10 后面插入节点 20 的时候，便可以先进行判断，节点 10 是否已经被删除，从而避免在一个删除的节点 10 后面插入节点 20。**这个解决方法有一个关键点：**"把节点 10 的 next 指针指向节点 20（插入操作）"和"判断节点 10 本身是否已经删除（判断操作）"，必须是原子的，必须在 1 个 CAS 操作里面完成！

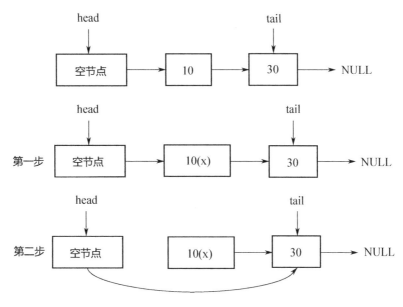

图 5-17 删除 1 个节点的两个步骤

具体的实现有两个办法：

办法一：AtomicMarkableReference

保证每个 next 是 AtomicMarkableReference 类型。但这个办法不够高效，Doug Lea 在 ConcurrentSkipListMap 的实现中用了另一种办法。

办法 2：Mark 节点

我们的目的是标记节点 10 已经删除，也就是标记它的 next 字段。那么可以新造一个 marker 节点，使节点 10 的 next 指针指向该 Marker 节点。这样，当向节点 10 的后面插入节点 20 的时候，就可以在插入的同时判断节点 10 的 next 指针是否执行了一个 Marker 节点，这两个操作可以在一个 CAS 操作里面完成。

3. 跳查表

解决了无锁链表的插入或删除问题，也就解决了跳查表的一个关键问题。因为跳查表就是多层链表叠起来的。

下面先看一下跳查表的数据结构（下面所用代码都引用自 JDK 7，JDK 8 中的代码略有差异，但不影响下面的原理分析）。

```
//底层 Node 节点。所有的<k,v>对，都是由这个单向链表串起来的
static final class Node<K,V> {
```

```
        final K key;
        volatile Object value;
        volatile Node<K,V> next;
        ...
    }
//上面的Index层的节点
    static class Index<K,V> {
        final Node<K,V> node;         //不存储实际数据,指向Node
        final Index<K,V> down;        //关键点:每个Index节点,必须有一个指针,指
                                      //向其下一个Level对应的节点
        volatile Index<K,V> right;    //自己也组成单向链表
        ...
    }
//整个ConcurrentSkipListMap就只需记录顶层的head节点
public class ConcurrentSkipListMap<K,V> extends AbstractMap<K,V>
    implements ConcurrentNavigableMap<K,V>,
               Cloneable,
               java.io.Serializable {
    ...
    private transient volatile HeadIndex<K,V> head;
}
```

跳查表的数据结构如图5-18所示。

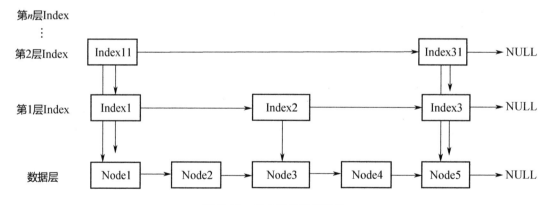

图5-18　跳查表的数据结构

下面详细分析如何从跳查表上查找、插入和删除元素。

（1）put实现分析。

```
public V put(K key, V value) {
    if (value == null)
        throw new NullPointerException();
```

```java
        return doPut(key, value, false);
    }
private V doPut(K kkey, V value, boolean onlyIfAbsent) {
        Comparable<? super K> key = comparable(kkey);
        for (;;) {
            Node<K,V> b = findPredecessor(key);  //要插入节点,需要先找到节点的前驱
Node<K,V> n = b.next;   //元素要插入在[b,n]之间
            for (;;) {
                if (n != null) {
                    Node<K,V> f = n.next;
                    if (n != b.next)                  //n != b.next,重新开始
                        break;
                    Object v = n.value;
                    if (v == null) {               //发现 n 已经被删除了,执行删除清理逻辑
                        n.helpDelete(b, f);
                        break;
                    }
                    if (v == n || b.value == null)   //发现 b 已经被删除了,重新开始
                        break;
                    int c = key.compareTo(n.key);
                    if (c > 0) {    //待插入的元素,已经大于n,则把[b,n]往后挪一个位置
                        b = n;
                        n = f;
                        continue;
                    }
                    if (c == 0) {       //节点存在,直接改值
                        if (onlyIfAbsent || n.casValue(v, value))
                            return (V)v;
                        else
                            break; //
                    }
                }
                //若待插入的元素在[b,n]之间,则会走到这个位置,插入新节点
                Node<K,V> z = new Node<K,V>(kkey, value, n);
                if (!b.casNext(n, z))   //若节点不存在,则新建一个,插入进去
                    break;

//判断是否要给此节点加上 Index 层。randomLevel 有 50%的概率会返回 0,也就是说,50%的节
//点不会有 Index 层,而只会在底层的链表上面
                int level = randomLevel();
                if (level > 0)
                    insertIndex(z, level);   //level > 0,为其建索引
```

```
            return null;
        }
    }
}
//关键函数：查找一个节点的前驱，从header开始，从左往右、从上往下遍历
    private Node<K,V> findPredecessor(Comparable<? super K> key) {
        if (key == null)
            throw new NullPointerException();
        for (;;) {
            Index<K,V> q = head;  //顶层index的头部
            Index<K,V> r = q.right;
            for (;;) {
                if (r != null) {
                    Node<K,V> n = r.node;
                    K k = n.key;
                    if (n.value == null) {  //r对应的Node已经是删除状态
                        if (!q.unlink(r))
                            break;
                        r = q.right;
                        continue;
                    }
                    if (key.compareTo(k) > 0) {  //要查找的元素 > 节点，一直往右遍历
                        q = r;
                        r = r.right;
                        continue;
                    }
                }
                //从上面的if判断中跳出来，则当前节点处于[q,r]之间，大于q，小于r，跳
                //到下一层，从q开始
Index<K,V> d = q.down;
                if (d != null) {
                    q = d;
                    r = d.right;
                } else   //到了底层，返回前驱
                    return q.node;
            }
        }
    }
```

如图5-19所示为跳查表查找示意图。

在底层，节点按照从小到大的顺序排列，上面的index层间隔地串在一起，因为从小到大排列。查找的时候，从顶层index开始，自左往右、自上往下，形成图示的遍历曲线。假设要查找

的元素是 32，遍历过程如下：

先遍历第 2 层 Index，发现在 21 的后面；

从 21 下降到第 1 层 Index，从 21 往后遍历，发现在 21 和 35 之间；

从 21 下降到底层，从 21 往后遍历，最终发现在 29 和 35 之间。

在整个的查找过程中，范围不断缩小，最终定位到底层的两个元素之间。

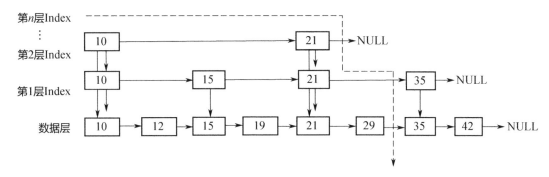

图 5-19 跳查表查找示意图

关于上面的 put(..)函数，有一个关键点需要说明：在通过 findPredecessor 找到了待插入的元素在[b,n]之间之后，并不能马上插入。因为其他线程也在操作这个链表，b、n 都有可能被删除，所以在插入之前执行了一系列的检查逻辑，而这也正是无锁链表的复杂之处。

（2）remove(..)分析

```
public V remove(Object key) {
return doRemove(key, null);
}
//若找到了(k,v)，则删除，并返回 v；若找不到，则返回 null
final V doRemove(Object okey, Object value) {
Comparable<? super K> key = comparable(okey);
for (;;) {
   Node<K,V> b = findPredecessor(key);
   Node<K,V> n = b.next;
   for (;;) {
      if (n == null)
         return null;
      Node<K,V> f = n.next;
      if (n != b.next)          //不一致读，重新开始
         break;
      Object v = n.value;
```

```
    if (v == null) {               //发现n已经删除, 执行删除的清理逻辑
        n.helpDelete(b, f);
        break;
    }
    if (v == n || b.value == null)   //发现b已经删除, 重新开始
        break;

    int c = key.compareTo(n.key);
    if (c < 0)         //要删除的元素小于n, 说明没找到要删除的元素, 返回null
        return null;
    if (c > 0) {       //要删除的元素大于n, [b,n]后移一个位置, 重新找
        b = n;
        n = f;
        continue;
    }
    //没找到要删除的元素(k,v)。key 相等, 但 value 不匹配, 返回null
    if (value != null && !value.equals(v))
            return null;
    //要删除的元素等于n。执行下面一系列的删除逻辑
    if (!n.casValue(v, null))
        break;
    if (!n.appendMarker(f) || !b.casNext(n, f))
        findNode(key);
    else {
        findPredecessor(key);
        if (head.right == null)
            tryReduceLevel();
    }
    return (V)v;
    }
  }
}
```

上面的删除函数和插入函数的逻辑非常类似, 因为无论是插入, 还是删除, 都要先找到元素的前驱, 也就是定位到元素所在的区间[b,n]。在定位之后, 执行下面几个步骤:

如果发现 b、n 已经被删除了, 则执行对应的删除清理逻辑;

否则, 如果没有找到待删除的(k,v), 返回 null;

如果找到了待删除的元素, 也就是节点 n, 则把 n 的 value 置为 null, 同时在 n 的后面加上 Marker 节点, 同时检查是否需要降低 Index 的层次。

（3）get 分析

```
public V get(Object key) {
    return doGet(key);
}
private V doGet(Object okey) {
    Comparable<? super K> key = comparable(okey);
    for (;;) {
        Node<K,V> n = findNode(key);
        if (n == null)
            return null;
        Object v = n.value;
        if (v != null)
            return (V)v;
    }
}
private Node<K,V> findNode(Comparable<? super K> key) {
    for (;;) {
        Node<K,V> b = findPredecessor(key);
        Node<K,V> n = b.next;   //待查找元素处于[b,n]之间
        for (;;) {
            if (n == null)
                return null;
            Node<K,V> f = n.next;
            //下面这段删除的清理逻辑和前面的一样
            if (n != b.next)
                break;
            Object v = n.value;
            if (v == null) {
                n.helpDelete(b, f);
                break;
            }
            if (v == n || b.value == null)
                break;
            int c = key.compareTo(n.key);
            if (c == 0)       //找到了，返回
                return n;
            if (c < 0)        //没有找到，返回 null
                return null;
            b = n;      //c > 0，后移 1 个位置，重新找
            n = f;
        }
```

            }
        }

通过上面的代码会发现,无论是插入、删除,还是查找,都有相似的逻辑,都需要先定位到元素位置[b,n],然后判断 b、n 是否已经被删除,如果是,则需要执行相应的删除清理逻辑。这也正是无锁链表复杂的地方。

5.6.2 ConcurrentSkipListSet

如下面代码所示,ConcurrentSkipListSet 只是对 ConcurrentSkipListMap 的简单封装,此处不再进一步展开叙述。

```java
public class ConcurrentSkipListSet<E>
extends AbstractSet<E>
implements NavigableSet<E>, Cloneable, java.io.Serializable {
    //封装了一个 ConcurrentSkipListMap
    private final ConcurrentNavigableMap<E,Object> m;
    public ConcurrentSkipListSet() {
        m = new ConcurrentSkipListMap<E,Object>();
    }
    public boolean add(E e) {
        return m.putIfAbsent(e, Boolean.TRUE) == null;
    }
    ...
}
```

第 6 章 线程池与 Future

6.1 线程池的实现原理

图 6-1 所示为线程池的实现原理:调用方不断地向线程池中提交任务;线程池中有一组线程,不断地从队列中取任务,这是一个典型的生产者—消费者模型。

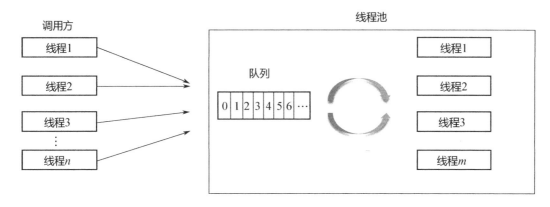

图 6-1 线程池的实现原理

原要实现这样一个线程池,有几个问题需要考虑:

(1)队列设置多长?如果是无界的,调用方不断地往队列中放任务,可能导致内存耗尽。如果是有界的,当队列满了之后,调用方如何处理?

(2)线程池中的线程个数是固定的,还是动态变化的?

(3)每次提交新任务,是放入队列?还是开新线程?

（4）当没有任务的时候，线程是睡眠一小段时间？还是进入阻塞？如果进入阻塞，如何唤醒？

针对问题（4），有3种做法：

做法（1）：不使用阻塞队列，只使用一般的线程安全的队列，也无阻塞—唤醒机制。当队列为空时，线程池中的线程只能睡眠一会儿，然后醒来去看队列中有没有新任务到来，如此不断轮询。

做法（2）：不使用阻塞队列，但在队列外部、线程池内部实现了阻塞—唤醒机制。

做法（3）：使用阻塞队列。

很显然，做法（3）最完善，既避免了线程池内部自己实现阻塞—唤醒机制的麻烦，也避免了做法（1）的睡眠—轮询带来的资源消耗和延迟。正因为如此，接下来要讲的ThreadPoolExecutor/ScheduledThreadPoolExecutor都是基于阻塞队列来实现的，而不是一般的队列，至此，各式各样的阻塞队列就要派上用场了。

6.2 线程池的类继承体系

在正式了解线程池的实现原理之前，先对线程池的类继承体系进行一个宏观介绍，如图6-2所示。

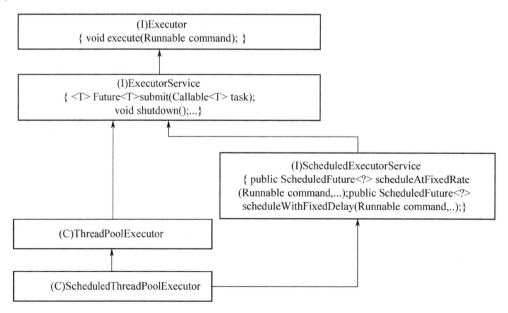

图6-2 ThreadPoolExecutor 和 ScheduledThreadPoolExecutor 类继承体系

在这里，有两个核心的类：ThreadPoolExecutor 和 ScheduledThreadPoolExecutor，后者不仅可以执行某个任务，还可以周期性地执行任务。

向线程池中提交的每个任务，都必须实现 Runnable 接口，通过最上面的 Executor 接口中的 execute(Runnable command)向线程池提交任务。

然后，在 ExecutorService 中，定义了线程池的关闭接口 shutdown()，还定义了可以有返回值的任务，也就是 Callable，接下来的章节会详细介绍。

6.3　ThreadPoolExecutor

6.3.1　核心数据结构

基于线程池的实现原理，下面看一下 ThreadPoolExecutor 的核心数据结构。

```
public class ThreadPoolExecutor extends AbstractExecutorService {
...
//状态变量（在接下来的章节中会详细介绍）
private final AtomicInteger ctl = new AtomicInteger(ctlOf(RUNNING, 0));
   private final BlockingQueue<Runnable> workQueue;   //存放任务的阻塞队列
   private final ReentrantLock mainLock = new ReentrantLock();
//对线程池内部各种变量进行互斥访问控制
   private final HashSet<Worker> workers = new HashSet<Worker>();   //线程集合
}
```

每一个线程是一个 Worker 对象。Worker 是 ThreadPoolExecutor 的内部类，核心数据结构如下：

```
private final class Worker extends AbstractQueuedSynchronizer
   implements Runnable
{
   ...
   final Thread thread;   //Worker 封装的线程
Runnable firstTask;   //Worker 接收到的第 1 个任务
volatile long completedTasks;   //Worker 执行完毕的任务个数
   }
```

由定义会发现，Worker 继承于 AQS，也就是说 Worker 本身就是一把锁。这把锁有什么用处呢？在接下来分析线程池的关闭、线程执行任务的过程时会了解到。

6.3.2　核心配置参数解释

针对本章最开始提出的线程池实现的几个问题，ThreadPoolExecutor 在其构造函数中提供了几个核心配置参数，来配置不同策略的线程池。了解了清楚每个参数的含义，也就明白了线程池的各种不同策略。

```java
public ThreadPoolExecutor(int corePoolSize,
                          int maximumPoolSize,
                          long keepAliveTime,
                          TimeUnit unit,
                          BlockingQueue<Runnable> workQueue,
                          ThreadFactory threadFactory,
                          RejectedExecutionHandler handler) {
    if (corePoolSize < 0 ||
        maximumPoolSize <= 0 ||
        maximumPoolSize < corePoolSize ||
        keepAliveTime < 0)
        throw new IllegalArgumentException();
    if (workQueue == null || threadFactory == null || handler == null)
        throw new NullPointerException();
    this.corePoolSize = corePoolSize;
    this.maximumPoolSize = maximumPoolSize;
    this.workQueue = workQueue;
    this.keepAliveTime = unit.toNanos(keepAliveTime);
    this.threadFactory = threadFactory;
    this.handler = handler;
}
```

上面的各个参数，解释如下：

（1）corePoolSize：在线程池中始终维护的线程个数。

（2）maxPoolSize：在 corePooSize 已满、队列也满的情况下，扩充线程至此值。

（3）keepAliveTime/TimeUnit：maxPoolSize 中的空闲线程，销毁所需要的时间，总线程数收缩回 corePoolSize。

（4）blockingQueue：线程池所用的队列类型。

（5）threadFactory：线程创建工厂，可以自定义，也有一个默认的。

（6）RejectedExecutionHandler：corePoolSize 已满，队列已满，maxPoolSize 已满，最后的拒绝策略。

下面来看这 6 个配置参数在任务的提交过程中是怎么运作的。在每次往线程池中提交任务的时候，有如下的处理流程：

step1：判断当前线程数是否大于或等于 corePoolSize。如果小于，则新建线程执行；如果大于，则进入 step2。

step2:判断队列是否已满。如未满,则放入;如已满,则进入 step3。

step3:判断当前线程数是否大于或等于 maxPoolSize。如果小于,则新建线程执行;如果大于,则进入 step4。

step4:根据拒绝策略,拒绝任务。

总结一下:首先判断 corePoolSize,其次判断 blockingQueue 是否已满,接着判断 maxPoolSize,最后使用拒绝策略。

很显然,基于这种流程,如果队列是无界的,将永远没有机会走到 step 3,也即 maxPoolSize 没有使用,也一定不会走到 step 4。

6.3.3 线程池的优雅关闭

在 1.1 章节中,讲了线程的优雅关闭,是一个很需要注意的地方。而线程池的关闭,较之线程的关闭更加复杂。当关闭一个线程池的时候,有的线程还正在执行某个任务,有的调用者正在向线程池提交任务,并且队列中可能还有未执行的任务。因此,关闭过程不可能是瞬时的,而是需要一个平滑的过渡,这就涉及线程池的完整生命周期管理。

1. 线程池的生命周期

在 JDK 7 中,把线程数量(workerCount)和线程池状态(runState)这两个变量打包存储在一个字段里面,即 ctl 变量。如图 6-3 所示,最高的 3 位存储线程池状态,其余 29 位存储线程个数。而在 JDK 6 中,这两个变量是分开存储的。

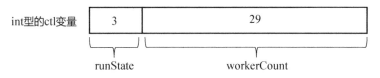

图 6-3 ctl 变量的 bit 位布局

```
//初始,线程池状态为 RUNNING,线程数为 0
    private final AtomicInteger ctl = new AtomicInteger(ctlOf(RUNNING, 0));
    //最高的 3 位表示线程池的状态
private static final int COUNT_BITS = Integer.SIZE - 3;
private static final int CAPACITY   = (1 << COUNT_BITS) - 1;

    //线程池的 5 种状态
private static final int RUNNING    = -1 << COUNT_BITS;
private static final int SHUTDOWN   =  0 << COUNT_BITS;
private static final int STOP       =  1 << COUNT_BITS;
```

```
private static final int TIDYING    = 2 << COUNT_BITS;
private static final int TERMINATED = 3 << COUNT_BITS;
    //从ctl中分别解包出runState和workerCount
    private static int runStateOf(int c)     { return c & ~CAPACITY; }
private static int workerCountOf(int c)  { return c & CAPACITY; }
    //rs即runState, wc即workerCount, 两个变量打包成ctl一个变量
    private static int ctlOf(int rs, int wc) { return rs | wc; }
```

由上面的代码可以看到，ctl 变量被拆成两半，最高的 3 位用来表示线程池的状态，低的 29 位表示线程的个数。线程池的状态有五种，分别是 RUNNING、SHUTDOWN、STOP、TIDYING 和 TERMINATED。

下面分析状态之间的迁移过程，如图 6-4 所示。

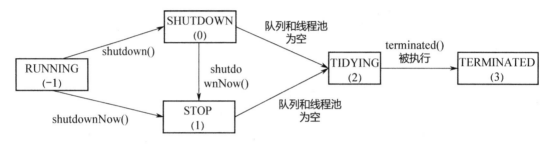

图 6-4 线程池的状态迁移图

线程池有两个关闭函数，shutdown()和 shutdownNow()，这两个函数会让线程池切换到不同的状态。在队列为空，线程池也为空之后，进入 TIDYING 状态；最后执行一个钩子函数 terminated()，进入 TERMINATED 状态，线程池才"寿终正寝"。

这里的状态迁移有一个非常关键的特征：从小到大迁移，-1，0，1，2，3，只会从小的状态值往大的状态值迁移，不会逆向迁移。例如，当线程池的状态在 TIDYING = 2 时，接下来只可能迁移到 TERMINATED = 3，不可能迁移回 STOP = 1 或者其他状态。

除 terminated()之外，线程池还提供了其他几个钩子函数，这些函数的实现都是空的。如果想实现自己的线程池，可以重写这几个函数。

```
    protected void beforeExecute(Thread t, Runnable r) { }
    protected void afterExecute(Runnable r, Throwable t) { }
    protected void terminated() { }
```

2. 正确关闭线程池的步骤

通过上面的分析，我们知道了线程池的关闭需要一个过程，在调用 shutDown()或者 shutdownNow()之后，线程池并不会立即关闭，接下来需要调用 awaitTermination 来等待线程池

关闭。关闭线程池的正确步骤如下:

```
executor.shutdown(); //或者
executor.shuwdownNow();
//调完上面的操作之后,再循环调用awaitTermination,等待线程池真正终止
try {
    boolean loop = true;
    do {     //等待所有任务完成
        loop = !executor.awaitTermination(2, TimeUnit.SECONDS);
        //阻塞,直到线程池里所有任务结束
    } while(loop);
} catch (InterruptedException e) {
    ...
}
```

awaitTermination(..)函数的内部实现很简单,如下所示。不断循环判断线程池是否到达了最终状态 TERMINATED,如果是,就返回;如果不是,则通过 termination 条件变量阻塞一段时间,"苏醒"之后,继续判断。

```
public boolean awaitTermination(long timeout, TimeUnit unit)
throws InterruptedException {
    long nanos = unit.toNanos(timeout);
    final ReentrantLock mainLock = this.mainLock;
    mainLock.lock();
    try {
        for (;;) {
            if (runStateAtLeast(ctl.get(), TERMINATED))   //判断状态是不是 TERMINATED
                return true;
            if (nanos <= 0)
                return false;
            nanos = termination.awaitNanos(nanos);
        }
    } finally {
        mainLock.unlock();
    }
}
```

3. shutdown()与shutdownNow()的区别

下面的代码展示了 shutdown()和 shutdownNow()的区别:

(1)前者不会清空任务队列,会等所有任务执行完成;后者会清空任务队列。

(2)前者只会中断空闲的线程,后者会中断所有线程。

```
public void shutdown() {
    final ReentrantLock mainLock = this.mainLock;
    mainLock.lock();
    try {
        checkShutdownAccess();          //检测是否有关闭线程池的权限
        advanceRunState(SHUTDOWN);      //把状态设置到 SHUTDOWN
        interruptIdleWorkers();         //只中断空闲的线程
        onShutdown();                   //钩子函数，是空的
    } finally {
        mainLock.unlock();
    }
    tryTerminate();
}

public List<Runnable> shutdownNow() {
    List<Runnable> tasks;
    final ReentrantLock mainLock = this.mainLock;
    mainLock.lock();
    try {
        checkShutdownAccess();
        advanceRunState(STOP);          //把状态设置到 STOP
        interruptWorkers();             //中断所有线程
        tasks = drainQueue();           //清空队列
    } finally {
        mainLock.unlock();
    }
    tryTerminate();
    return tasks;
}
```

下面看一下在上面的代码里中断空闲线程和中断所有线程的区别。

```
private void interruptIdleWorkers(boolean onlyOne) {
    final ReentrantLock mainLock = this.mainLock;
    mainLock.lock();
    try {
        for (Worker w : workers) {
            Thread t = w.thread;
            if (!t.isInterrupted() && w.tryLock()) {   //关键点：tryLock 调用成功，
                                                       //说明线程处于空闲状态；tryLock
                                                       //调用不成功，说明线程当前持有
                                                       //锁，正在执行某个任务
                try {
                    t.interrupt();
```

```
                } catch (SecurityException ignore) {
                } finally {
                    w.unlock();
                }
            }
            if (onlyOne)
                break;
        }
    } finally {
        mainLock.unlock();
    }
}
private void interruptWorkers() {
    final ReentrantLock mainLock = this.mainLock;
    mainLock.lock();
    try {
        for (Worker w : workers)
            w.interruptIfStarted();     //不管线程是否正在执行任务,一律发送中断信号
    } finally {
        mainLock.unlock();
    }
}
```

关键区别点在 tryLock(): 一个线程在执行一个任务之前, 会先加锁(在 6.3.4 节会详细讲), 这意味着通过是否持有锁, 可以判断出线程是否处于空闲状态。tryLock()如果调用成功, 说明线程处于空闲状态, 向其发送中断信号; 否则不发送。

在上面的代码中, shutdown() 和 shutdownNow()都调用了 tryTerminate()函数, 如下所示。

```
final void tryTerminate() {
    for (;;) {
        int c = ctl.get();
        if (isRunning(c) ||
            runStateAtLeast(c, TIDYING) ||
            (runStateOf(c) == SHUTDOWN && ! workQueue.isEmpty()))
            return;
        if (workerCountOf(c) != 0) {
            interruptIdleWorkers(ONLY_ONE);
            return;
        }

        //当workQueue为空, workCount 为 0 时, 才会到这里
        final ReentrantLock mainLock = this.mainLock;
```

```
        mainLock.lock();
        try {
            if (ctl.compareAndSet(c, ctlOf(TIDYING, 0))) {   //把状态切换到 TIDYING
                try {
                    terminated();    //调用钩子函数
                } finally {
                    ctl.set(ctlOf(TERMINATED, 0));  //把状态由 TIDYING 改为 TERMINATED
                    termination.signalAll();          //通知 awaitTermination(..)
                }
                return;
            }
        } finally {
            mainLock.unlock();
        }
    }
}
```

tryTerminate() 不会强行终止线程池，只是做了一下检测：当 workerCount 为 0，workerQueue 为空时，先把状态切换到 TIDYING，然后调用钩子函数 terminated()。当钩子函数执行完成时，把状态从 TIDYING 改为 TERMINATED，接着调用 termination.sinaglAll()，通知前面阻塞在 awaitTermination 的所有调用者线程。

所以，TIDYING 和 TREMINATED 的区别是在二者之间执行了一个钩子函数 terminated()，目前是一个空实现。

6.3.4 任务的提交过程分析

提交任务的函数如下：

```
public void execute(Runnable command) {
if (command == null)
    throw new NullPointerException();
int c = ctl.get();
//如果当前的线程数小于 corePoolSize，则开新线程
if (workerCountOf(c) < corePoolSize) {
    if (addWorker(command, true))
        return;
    c = ctl.get();
}
//如果当前的线程数大于或等于 corePoreSize，则调用 workQueue.offer 放入队列
if (isRunning(c) && workQueue.offer(command)) {
    int recheck = ctl.get();
```

```
        if (! isRunning(recheck) && remove(command))
            reject(command);
        else if (workerCountOf(recheck) == 0)
            addWorker(null, false);
    } //放入队列失败,开新线程
    else if (!addWorker(command, false))
        reject(command);    //线程数大于maxPoolSize,调用拒绝策略
}
//此函数用于开一个新线程。如果第 2 个参数 core 为 true,则用 corePoolSize 作为上界;如果
//为 false,则用 maxPoolSize 作为上界
private boolean addWorker(Runnable firstTask, boolean core) {
    retry:
    for (;;) {
        int c = ctl.get();
        int rs = runStateOf(c);

        //只要状态大于或等于 SHUTDOWN,说明线程池进入了关闭的过程
        if (rs >= SHUTDOWN &&
            ! (rs == SHUTDOWN &&
               firstTask == null &&
               ! workQueue.isEmpty()))
            return false;

        for (;;) {
            int wc = workerCountOf(c);
            if (wc >= CAPACITY ||
                wc >= (core ? corePoolSize : maximumPoolSize))
                return false;    //线程数超过上界 corePoolSize 或者 maxPoolSize,不会开
                                 //新线程,直接返回 false

          if (compareAndIncrementWorkerCount(c)) //workCount 成功加 1,跳出整个 for 循环
                break retry;
            c = ctl.get();
            if (runStateOf(c) != rs) //runState 在这个过程中发生了变化,重新开始 for 循环
                continue retry;
        }
    }

    //workCount 成功加 1,开始添加线程操作
    boolean workerStarted = false;
    boolean workerAdded = false;
    Worker w = null;
```

```
try {
    final ReentrantLock mainLock = this.mainLock;
    w = new Worker(firstTask);    //创建一个线程
    final Thread t = w.thread;
    if (t != null) {
        mainLock.lock();
        try {
            int c = ctl.get();
            int rs = runStateOf(c);

            if (rs < SHUTDOWN ||
                (rs == SHUTDOWN && firstTask == null)) {
                if (t.isAlive())
                    throw new IllegalThreadStateException();
                workers.add(w);   // 把线程加入线程集合
                int s = workers.size();
                if (s > largestPoolSize)
                    largestPoolSize = s;
                workerAdded = true;
            }
        } finally {
            mainLock.unlock();
        }
        if (workerAdded) {
            t.start();    //若成功加入，则启动该线程
            workerStarted = true;
        }
    }
} finally {
    if (! workerStarted)    //加入失败，调用下面的函数
        addWorkerFailed(w);  //在这个函数内部会把workCount 减 1
}
return workerStarted;
}
```

6.3.5 任务的执行过程分析

在上面的任务提交过程中，可能会开启一个新的 Worker，并把任务本身作为 firstTask 赋给该 Worker。但对于一个 Worker 来说，不是只执行一个任务，而是源源不断地从队列中取任务执行，这是一个不断循环的过程。

下面来看 Woker 的 run()方法的实现过程。

```
private final class Worker extends AbstractQueuedSynchronizer
implements Runnable{
Worker(Runnable firstTask) {
  setState(-1);   //初始状态是-1
this.firstTask = firstTask;
  this.thread = getThreadFactory().newThread(this);
}
public void run() {
  runWorker(this);    //调用了ThreadPoolExecutor 的 runWoker(Worker w)函数
}
}
//核心函数，ThreadPoolExecutor 的 runWorker(Worker w)
final void runWorker(Worker w) {
Thread wt = Thread.currentThread();
Runnable task = w.firstTask;
w.firstTask = null;
w.unlock();
    boolean completedAbruptly = true;
try {
    while (task != null || (task = getTask()) != null) {      //不断从队列中取
                                                              //任务执行
        w.lock();       //关键点：在执行任务之前要先加锁。此处就对应了前面讲 shutdown()
                        //时的 tryLock
        if ((runStateAtLeast(ctl.get(), STOP) ||
          (Thread.interrupted() &&
            runStateAtLeast(ctl.get(), STOP))) &&
          !wt.isInterrupted())
            wt.interrupt();     //拿到任务了，在执行之前重新检测线程池的状态。如果发现
                                //已经开始关闭，自己给自己发中断信号
        try {
            beforeExecute(wt, task);    //任务之前的钩子函数，目前是空实现
            Throwable thrown = null;
            try {
                task.run();     //执行任务代码
            } catch (RuntimeException x) {
                thrown = x; throw x;
            } catch (Error x) {
                thrown = x; throw x;
            } catch (Throwable x) {
                thrown = x; throw new Error(x);
            } finally {
                afterExecute(task, thrown);     //任务之后的钩子函数，目前是空实现
```

```
            }
        } finally {
            task = null;
            w.completedTasks++;  //成功完成任务，completedTasks 累加
            w.unlock();  //释放锁
        }
    }
    completedAbruptly = false;   //判断这个 Worker 是正常退出，还是收到中断退出，或
                                 //者其他某种异常退出
} finally {
    processWorkerExit(w, completedAbruptly);  //Worker 退出
}
}
```

1. shutdown()与任务执行过程综合分析

把任务的执行过程和上面的线程池的关闭过程结合起来进行分析，当调用 shutdown()的时候，可能出现以下几种场景：

场景 1：当调用 shutdown()的时候，所有线程都处于空闲状态。

这意味着任务队列一定是空的。此时，所有线程都会阻塞在 getTask()函数的地方。然后，所有线程都会收到 interruptIdleWorkers()发来的中断信号，getTask()返回 null，所有 Worker 都会退出 while 循环，之后执行 processWorkerExit。

场景 2：当调用 shutdown()的时候，所有线程都处于忙碌状态。

此时，队列可能是空的，也可能是非空的。interruptIdleWorkers()内部的 tryLock 调用失败，什么都不会做，所有线程会继续执行自己当前的任务。之后所有线程会执行完队列中的任务，直到队列为空，getTask()才会返回 null。之后，就和场景 1 一样了，退出 while 循环。

场景 3：当调用 shutdown()的时候，部分线程忙碌，部分线程空闲。

有部分线程空闲，说明队列一定是空的，这些线程肯定阻塞在 getTask()函数的地方。空闲的这些线程会和场景 1 一样处理，不空闲的线程会和场景 2 一样处理。

下面看一下 getTask()函数的内部细节。

```
private Runnable getTask() {
boolean timedOut = false;

retry:
for (;;) {
    int c = ctl.get();
    int rs = runStateOf(c);
```

```
    //关键点：
       //1.如果 rs >= STOP，即调用了 shutdownNow()，此处会返回 null
       //2.如果 rs >= SHUTDOWN，即调用了 shutdown()，并且队列为空，此处也会返回 null
    if (rs >= SHUTDOWN && (rs >= STOP || workQueue.isEmpty())) {
        decrementWorkerCount();
        return null;   //此处返回 null，上面的 Worker 就会退出 while 循环，死亡
    }

    boolean timed;

    for (;;) {
        int wc = workerCountOf(c);
        timed = allowCoreThreadTimeOut || wc > corePoolSize;

        if (wc <= maximumPoolSize && ! (timedOut && timed))
            break;
        if (compareAndDecrementWorkerCount(c))
            return null;
        c = ctl.get();
        if (runStateOf(c) != rs)
            continue retry;
    }

    try {
//关键点：
//1.队列为空，就会阻塞此处的 poll 或者 take，线程空闲。前置带超时，后置不带超时
//2.一旦收到中断信号，此处就会抛出中断异常。对应上文的场景 1
        Runnable r = timed ?
            workQueue.poll(keepAliveTime, TimeUnit.NANOSECONDS) :
            workQueue.take();
        if (r != null)
            return r;
        timedOut = true;
    } catch (InterruptedException retry) {
        timedOut = false;
    }
}
}
```

2. shutdownNow() 与任务执行过程综合分析

和上面的 shutdown() 类似，只是多了一个环节，即清空任务队列。在第 1 章中已经讲到，

如果一个线程正在执行某个业务代码，即使向它发送中断信号，也没有用，只能等它把代码执行完成。从这个意义上讲，中断空闲线程和中断所有线程的区别并不是很大，除非线程当前刚好阻塞在某个地方。

有兴趣的读者也可以分析在不同场景下调用 shutdownNow() 所遇到的情况，此处不再展开。

当一个 Worker 最终退出的时候，会执行清理工作，代码如下所示。

```
private void processWorkerExit(Worker w, boolean completedAbruptly) {
//如果一个 Woker 正常退出，在上面的 getTask() 里面，就已经把 workerCount 减 1。走到这个地
//方，都是非正常退出（例如执行某个业务逻辑的时候报错了），要在这个地方，把 workerCount 减 1
  if (completedAbruptly)
      decrementWorkerCount();

final ReentrantLock mainLock = this.mainLock;
mainLock.lock();
try {
    completedTaskCount += w.completedTasks;
    workers.remove(w);     //把自己从 wokers 集合中移除
} finally {
    mainLock.unlock();
}

//和 shutdown/shutdownNow() 一样，每个线程在结束的时候都会尝试调用这个函数，看是否可
//以终止整个线程池
tryTerminate();

//最后这一段是一个保险：自己要退出的时候，发现线程池的状态小于 STOP，并且队列不为空，
//并且当前没有工作线程数了，那么调用 addWorker() 再开启一个新线程，把队列中的任务消耗掉
int c = ctl.get();
if (runStateLessThan(c, STOP)) {
    if (!completedAbruptly) {
        int min = allowCoreThreadTimeOut ? 0 : corePoolSize;
        if (min == 0 && ! workQueue.isEmpty())
            min = 1;
        if (workerCountOf(c) >= min)
            return; // replacement not needed
    }
    addWorker(null, false);
}
}
```

6.3.6 线程池的 4 种拒绝策略

在 execute（Runnable command）的最后，调用了 reject（command）执行拒绝策略，代码如下所示。

```
final void reject(Runnable command) {
handler.rejectedExecution(command, this);
}
    private volatile RejectedExecutionHandler handler;
```

RejectedExecutionHandler 是一个接口，定义了四种实现，分别对应四种不同的拒绝策略，默认是 AbortPolicy。四种策略的实现代码如下：

```
//策略1：让调用者直接在自己的线程里面执行，线程池不做处理
    public static class CallerRunsPolicy implements RejectedExecutionHandler {
        public CallerRunsPolicy() { }
        public void rejectedExecution(Runnable r, ThreadPoolExecutor e) {
            if (!e.isShutdown()) {
                r.run();
            }
        }
    }
//策略2：线程池直接抛出异常
    public static class AbortPolicy implements RejectedExecutionHandler {
        public AbortPolicy() { }
        public void rejectedExecution(Runnable r, ThreadPoolExecutor e) {
            throw new RejectedExecutionException("Task " + r.toString() +
                                                 " rejected from " +
                                                 e.toString());
        }
    }

//策略3：线程池直接把任务丢掉，当作什么也没发生
    public static class DiscardPolicy implements RejectedExecutionHandler {
        public DiscardPolicy() { }
        public void rejectedExecution(Runnable r, ThreadPoolExecutor e) {
        }
    }
//策略4：把队列里面最老的任务删除掉，把该任务放入队列中
    public static class DiscardOldestPolicy implements RejectedExecutionHandler {
        public DiscardOldestPolicy() { }
        public void rejectedExecution(Runnable r, ThreadPoolExecutor e) {
            if (!e.isShutdown()) {
                e.getQueue().poll();
```

```
            e.execute(r);
        }
    }
}
```

6.4 Callable 与 Future

execute(Runnable command)接口是无返回值的,与之相对应的是一个有返回值的接口 Future submit(Callable task),这点在 6.2 节中已经提到。

Callable 也就是一个有返回值的 Runnable,其定义如下所示。

```
public interface Callable<V> {
    V call() throws Exception;
}
```

使用方式如下:

```
//自定义一个Callable,类似定义一个Runnable
Callable<String> c = new XXXCallable<String>();
//把Callable提交给线程池执行,线程池会立马返回一个Future,"票据"
Future<String> f = executor.submit(c);
//通过"票据"取回结果。如果任务没有计算完,调用者一直阻塞在这里
String result = f.get();
//Future接口的定义如下,主要用来取回将来要返回的结果
public interface Future<V> {
...
    V get() throws InterruptedException, ExecutionException;
    V get(long timeout, TimeUnit unit)
        throws InterruptedException, ExecutionException, TimeoutException;
}
```

submit(Callable task)并不是在 ThreadPoolExecutor 里面直接实现的,而是实现在其父类 AbstractExecutorService 中,源码如下:

```
public abstract class AbstractExecutorService implements ExecutorService {
...
public <T> Future<T> submit(Callable<T> task) {
if (task == null) throw new NullPointerException();
RunnableFuture<T> ftask = newTaskFor(task);  //把Callable转换成Runnable
execute(ftask);   //调用ThreadPoolExecutor的execute(Runnable command)接口
return ftask;
}
```

从这段代码中可以看出，Callable 其实是用 Runnable 实现的。在 submit 内部，把 Callable 通过 FutureTask 这个 Adapter 转化成 Runnable，然后通过 execute 执行。如图 6-5 所示为 Callable 被转换成 Runnable 示意图。

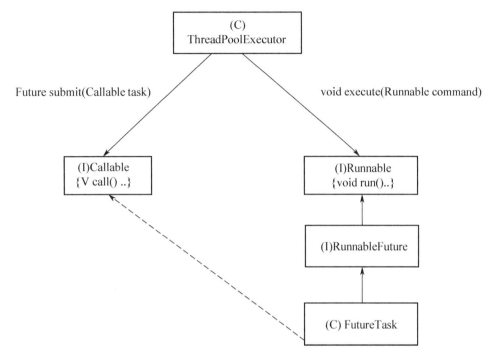

图 6-5　Callable 被转换成 Runnable 示意图

FutureTask 是一个 Adapter 对象。一方面，它实现了 Runnable 接口，也实现了 Future 接口；另一方面，它的内部包含了一个 Callable 对象，从而实现了把 Callable 转换成 Runnable。

```
public interface RunnableFuture<V> extends Runnable, Future<V> {
    void run();
}
//实现了 Runnable 接口、Future 接口
public class FutureTask<V> implements RunnableFuture<V> {
...
private Callable<V> callable;  //内部封装的 Callable 对象
private Object outcome;    //Callable 的执行结果
private volatile int state;  //CAS state 变量 + LockSupport.park/unpark
public FutureTask(Callable<V> callable) {
if (callable == null)
    throw new NullPointerException();
this.callable = callable;
```

```java
        this.state = NEW;
    }


    public void run() {
        if (state != NEW ||
            !UNSAFE.compareAndSwapObject(this, runnerOffset,
                                        null, Thread.currentThread()))
            return;
        try {
            Callable<V> c = callable;
            if (c != null && state == NEW) {
                V result;
                boolean ran;
                try {
//关键：把 Callable 的 call()转换成 Runnable 的 run()
                    result = c.call();
                    ran = true;
                } catch (Throwable ex) {
                    result = null;
                    ran = false;
                    setException(ex);
                }
                if (ran)
                    set(result);    //把返回值存入 outcome 变量
            }
        } finally {
            runner = null;
            int s = state;
            if (s >= INTERRUPTING)
                handlePossibleCancellationInterrupt(s);
        }
    }
    public V get() throws InterruptedException, ExecutionException {
        int s = state;
        if (s <= COMPLETING)
            s = awaitDone(false, 0L);   //内部调用了 LockSupport.park()函数
        return report(s);
    }
}
```

如图 6-6 所示，一方面，线程池内部的线程在执行 RunTask 的 run()方法；另一方面，外部

多个线程又在调用 get()方法，等着返回结果，因此这个地方需要一个阻塞—通知机制。

图 6-6　FutureTask 对象的线程同步示意图

在 JDK 6 中借用 AQS 的功能来实现阻塞—唤醒。但自 JDK 7 开始，既没有借用 AQS 的功能，也没有使用 Condition 的 await()/notify()机制，而是直接基于 CAS state 变量 + park/unpark()来实现阻塞—唤醒机制。由于这个原理在上文讲 AQS 和 Condition 的时候已反复提及，此处就不再对 awaitDone()进一步展开分析了。

6.5　ScheduledThreadPoolExecutor

ScheduledThreadPoolExecutor 实现了按时间调度来执行任务，具体而言有两个方面：

（1）延迟执行任务

```
public ScheduledFuture<?> schedule(Runnable command, long delay, TimeUnit unit);
public <V> ScheduledFuture<V> schedule(Callable<V> callable,long delay,TimeUnit unit);
```

（2）周期执行任务

```
public ScheduledFuture<?> scheduleAtFixedRate(Runnable command,long initialDelay, long period,TimeUnit unit);
public ScheduledFuture<?> scheduleWithFixedDelay(Runnable command,long initialDelay,
long delay,TimeUnit unit);
```

这两个函数的区别如下：

AtFixedRate：按固定频率执行，与任务本身执行时间无关。但有个前提条件，任务执行时间必须小于间隔时间，例如间隔时间是 5s，每 5s 执行一次任务，任务的执行时间必须小于 5s。

WithFixedDelay：按固定间隔执行，与任务本身执行时间有关。例如，任务本身执行时间是 10s，间隔 2s，则下一次开始执行的时间就是 12s。

6.5.1 延迟执行和周期性执行的原理

从 6.2 节的类继承体系可以看到，ScheduledThreadPoolExecutor 继承了 ThreadPoolExecutor，这意味着其内部的数据结构和 ThreadPoolExecutor 是基本一样的，那它是如何实现延迟执行任务和周期性执行任务的呢？

延迟执行任务依靠的是 DelayQueue。在 5.1 节中已经提到，DelayQueue 是 BlockingQueue 的一种，其实现原理是二叉堆。

而周期性执行任务是执行完一个任务之后，再把该任务扔回到任务队列中，如此就可以对一个任务反复执行。

不过这里并没有使用 5.1 节中的 DelayQueue，而是在 ScheduledThreadPoolExecutor 内部又实现了一个特定的 DelayQueue，如下所示。

```
//ScheduledThreadPoolExecutor 的内部类
static class DelayedWorkQueue extends AbstractQueue<Runnable>
    implements BlockingQueue<Runnable> {
    ...
}
```

其原理和 DelayQueue 一样，但针对任务的取消进行了优化，此处不再进一步展开。下面主要看一下延迟执行和周期性执行的实现过程。

6.5.2 延迟执行

```
public ScheduledFuture<?> schedule(Runnable command, long delay, TimeUnit unit) {
    if (command == null || unit == null)
        throw new NullPointerException();

    RunnableScheduledFuture<?> t = decorateTask(command,
        new ScheduledFutureTask<Void>(command, null,
                        triggerTime(delay, unit)));
    delayedExecute(t);
    return t;
}
```

传进去的是一个 Runnable，外加延迟时间 delay。在内部通过 decorateTask(..) 函数把 Runnable 包装成一个 ScheduleFutureTask 对象，而 DelayedWorkerQueue 中存放的正是这种类型的对象，这种类型的对象一定实现了 Delayed 接口，如 5.1.4 节所述。

```
private void delayedExecute(RunnableScheduledFuture<?> task) {
    if (isShutdown())
```

```
        reject(task);
    else {
        super.getQueue().add(task);  //把 ScheduleFutureTask 对象加入延迟队列
        if (isShutdown() &&
            !canRunInCurrentRunState(task.isPeriodic()) &&
            remove(task))
            task.cancel(false);
        else
            ensurePrestart();
    }
}
//如果 wc < corePoolSize，则会开新线程。否则什么都不做
void ensurePrestart() {
    int wc = workerCountOf(ctl.get());
    if (wc < corePoolSize)
        addWorker(null, true);
    else if (wc == 0)
        addWorker(null, false);
}
```

从上面的代码中可以看出，schedule()函数本身很简单，就是把提交的 Runnable 任务加上 delay 时间，转换成 ScheduledFutureTask 对象，放入 DelayedWorkerQueue 中。任务的执行过程还是复用的 ThreadPoolExecutor，延迟的控制是在 DelayedWorkerQueue 内部完成的。

6.5.3 周期性执行

```
    public ScheduledFuture<?> scheduleWithFixedDelay(Runnable command, long initialDelay,
long delay, TimeUnit unit) {
        if (command == null || unit == null)
            throw new NullPointerException();
        if (delay <= 0)
            throw new IllegalArgumentException();
        ScheduledFutureTask<Void> sft =
            new ScheduledFutureTask<Void>(command,
                                  null,
                                  triggerTime(initialDelay, unit),
                                  unit.toNanos(-delay));  //负数
        RunnableScheduledFuture<Void> t = decorateTask(command, sft);
        sft.outerTask = t;
        delayedExecute(t);
        return t;
```

```java
        }
        public ScheduledFuture<?> scheduleAtFixedRate(Runnable command,long initialDelay,
long period, TimeUnit unit) {
            if (command == null || unit == null)
                throw new NullPointerException();
            if (period <= 0)
                throw new IllegalArgumentException();
            ScheduledFutureTask<Void> sft =
                new ScheduledFutureTask<Void>(command,
                                    null,
                                    triggerTime(initialDelay, unit),
                                    unit.toNanos(period));   //正数
            RunnableScheduledFuture<Void> t = decorateTask(command, sft);
            sft.outerTask = t;
            delayedExecute(t);
            return t;
        }
```

和 schedule(..)函数的框架基本一样,也是包装一个 ScheduledFutureTask 对象,只是在延迟时间参数之外多了一个周期参数,然后放入 DelayedWorkerQueue 就结束了。

两个函数的区别在于一个传入的周期是一个负数,另一个传入的周期是一个正数,为什么要这样做呢?下面进入 ScheduledFutureTask 的内部一探究竟。

```java
    private class ScheduledFutureTask<V>
        extends FutureTask<V> implements RunnableScheduledFuture<V> {
    ...
private final long sequenceNumber;
private long time;
    private final long period;
    ScheduledFutureTask(Runnable r, V result, long ns, long period) {
    super(r, result);
    this.time = ns;   //延迟时间
    this.period = period;  //周期
    this.sequenceNumber = sequencer.getAndIncrement();   //自增序列号
}
    //实现 Delayed 接口
    public long getDelay(TimeUnit unit) {
    return unit.convert(time - now(), TimeUnit.NANOSECONDS);
}

    //实现 Comparable 接口
```

```java
public int compareTo(Delayed other) {
    if (other == this)
        return 0;
    if (other instanceof ScheduledFutureTask) {
        ScheduledFutureTask<?> x = (ScheduledFutureTask<?>)other;
        long diff = time - x.time;
        if (diff < 0)
            return -1;
        else if (diff > 0)
            return 1;
        else if (sequenceNumber < x.sequenceNumber)   //两个延迟时间相等，再进一
                                                      //步比较序列号
            return -1;
        else
            return 1;
    }
    long d = (getDelay(TimeUnit.NANOSECONDS) -
            other.getDelay(TimeUnit.NANOSECONDS));
    return (d == 0) ? 0 : ((d < 0) ? -1 : 1);
}
//实现Runnable接口
public void run() {
    boolean periodic = isPeriodic();
    if (!canRunInCurrentRunState(periodic))
        cancel(false);
    else if (!periodic)    //非周期的，执行一次就结束了
        ScheduledFutureTask.super.run();
    else if (ScheduledFutureTask.super.runAndReset()) {
    //周期的，执行完之后重新计算延迟时间，再扔回到队列
        setNextRunTime();
        reExecutePeriodic(outerTask);
    }
}
//设置下一次执行时间
private void setNextRunTime() {
    long p = period;
    if (p > 0)
        time += p;
    else
        time = triggerTime(-p);
}
```

```
        long triggerTime(long delay) {
        return now() +
            ((delay < (Long.MAX_VALUE >> 1)) ? delay : overflowFree(delay));
}
        //回到队列中,等待下一次执行
        void reExecutePeriodic(RunnableScheduledFuture<?> task) {
        if (canRunInCurrentRunState(true)) {
            super.getQueue().add(task);
            if (!canRunInCurrentRunState(true) && remove(task))
                task.cancel(false);
            else
                ensurePrestart();
        }
}
```

withFixedDelay 和 atFixedRate 的区别就体现在 setNextRunTime 里面。

如果是 atFixedRate,period > 0,下一次开始执行时间等于上一次开始执行时间 + period;

如果是 withFixedDelay,period < 0,下一次开始执行时间等于 triggerTime(-p),为 now + (-period),now 即上一次执行的结束时间。

6.6 Executors 工具类

concurrent 包提供了 Executors 工具类,利用它可以创建各种不同类型的线程池,如下所示。

```
public class Executors {
//单线程线程池
public static ExecutorService newSingleThreadExecutor() {
    return new FinalizableDelegatedExecutorService
        (new ThreadPoolExecutor(1, 1,
                                0L, TimeUnit.MILLISECONDS,
                                new LinkedBlockingQueue<Runnable>()));
}
//固定数目线程的线程池
public static ExecutorService newFixedThreadPool(int nThreads) {
    return new ThreadPoolExecutor(nThreads, nThreads,
                                  0L, TimeUnit.MILLISECONDS,
                                  new LinkedBlockingQueue<Runnable>());
    }
//CachedThreadPool,每来一个任务,就创建一个线程
public static ExecutorService newCachedThreadPool() {
```

```
        return new ThreadPoolExecutor(0, Integer.MAX_VALUE,
                               60L, TimeUnit.SECONDS,
                               new SynchronousQueue<Runnable>());
    }
    //单线程的,具有周期调度功能的线程池
     public static ScheduledExecutorService newSingleThreadScheduledExecutor() {
        return new DelegatedScheduledExecutorService
            (new ScheduledThreadPoolExecutor(1));
    }
    //多线程的,具有周期调度功能的线程池
    public    static    ScheduledExecutorService    newScheduledThreadPool(int corePoolSize) {
        return new ScheduledThreadPoolExecutor(corePoolSize);
    }
    ...
}
```

从上面的代码中可以看出,这些不同类型的线程池,其实都是由前面的几个关键配置参数配置而成的。

在《阿里巴巴 Java 开发手册》中,明确禁止使用 Executors 创建线程池,并要求开发者直接使用 ThreadPoolExecutor 或 ScheduledThreadPoolExecutor 进行创建。这样做是为了强制开发者明确线程池的运行策略,使其对线程池的每个配置参数皆做到心中有数,以规避因使用不当而造成资源耗尽的风险。

第 7 章
ForkJoinPool

7.1 ForkJoinPool 用法

在大学的算法课本中,都有一种基本算法:分治。其基本思路是:将一个大的任务分为若干个子任务,这些子任务分别计算,然后合并出最终结果,在这个过程中通常会用到递归。

而 ForkJoinPool 就是 JDK7 提供的一种"分治算法"的多线程并行计算框架。Fork 意为分叉,Join 意为合并,一分一合,相互配合,形成分治算法。此外,也可以将 ForkJoinPool 看作一个单机版的 Map/Reduce,只不过这里的并行不是多台机器并行计算,而是多个线程并行计算。

相比于 ThreadPoolExecutor,ForkJoinPool 可以更好地实现计算的负载均衡,提高资源利用率。假设有 5 个任务,在 ThreadPoolExecutor 中有 5 个线程并行执行,其中一个任务的计算量很大,其余 4 个任务的计算量很小,这会导致 1 个线程很忙,其他 4 个线程则处于空闲状态。而利用 ForkJoinPool,可以把大的任务拆分成很多小任务,然后这些小任务被所有的线程执行,从而实现任务计算的负载均衡。

下面通过两个简单例子看一下 ForkJoinPool 的用法:

例子 1:快排

快排有 2 个步骤:

第 1 步,利用数组的第 1 个元素把数组划分成两半,左边数组里面的元素小于或等于该元素,右边数组里面的元素比该元素大;

第 2 步,对左右的 2 个子数组分别排序。

可以看出,这里左右 2 个子数组是可以相互独立、并行计算的。因此可以利用 ForkJoinPool,

代码如下所示。

```java
//定义一个 Task,继承自 RecursiveAction,实现其 compute 方法
class SortTask extends RecursiveAction {
    final long[] array;
    final int lo;
    final int hi;
    private int THRESHOLD = 0; //For demo only
    public SortTask(long[] array) {
        this.array = array;
        this.lo = 0;
        this.hi = array.length - 1;
    }
    public SortTask(long[] array, int lo, int hi) {
        this.array = array;
        this.lo = lo;
        this.hi = hi;
    }
    protected void compute() {
      if(lo < hi) {
            int pivot = partition(array, lo, hi);  //划分
            SortTask left = new SortTask(array, lo, pivot - 1);
            SortTask right = new SortTask(array, pivot + 1, hi);
            left.fork();
            right.fork();
            left.join();
            right.join();
        }
    }
    private int partition(long[] array, int lo, int hi) {
        long x = array[hi];
        int i = lo - 1;
        for (int j = lo; j < hi; j++) {
            if (array[j] <= x) {
                i++;
                swap(array, i, j);
            }
        }
        swap(array, i + 1, hi);
        return i + 1;
    }
    private void swap(long[] array, int i, int j) {
```

```java
            if (i != j) {
                long temp = array[i];
                array[i] = array[j];
                array[j] = temp;
            }
        }
    }
}
//测试程序
public void testSort() throws Exception {
    ForkJoinTask sort = new SortTask(array);    //1个任务
    ForkJoinPool fjpool = new ForkJoinPool();    //1个ForkJoinPool
    fjpool.submit(sort);   //提交任务
    fjpool.shutdown();        //结束。ForkJoinPool 内部会开多个线程, 同时执行上面的子任务
    fjpool.awaitTermination(30, TimeUnit.SECONDS);
}
```

例子2：求1到 n 个数的和

```java
//定义一个Task, 继承自 RecursiveTask, 实现其 compute 方法
public class SumTask extends RecursiveTask<Long>{
    private static final int THRESHOLD = 10;
    private long start;
    private long end;
    public SumTask(long n) {
        this(1,n);
    }
    private SumTask(long start, long end) {
        this.start = start;
        this.end = end;
    }
    protected Long compute() {
        long sum = 0;
        if((end - start) <= THRESHOLD){
            for(long l = start; l <= end; l++){
                sum += l;
            }
        }else{
            long mid = (start + end) >>> 1;
            SumTask left = new SumTask(start, mid);     //分治, 递归
            SumTask right = new SumTask(mid + 1, end);
            left.fork();
            right.fork();
            sum = left.join() + right.join();
```

```
        }
        return sum;
    }
}
//测试函数
public void testSum() throws Exception {
SumTask sum = new SumTask(100);    //1个任务
ForkJoinPool fjpool = new ForkJoinPool();   //1个 ForkJoinPool
Future<Long> future = fjpool.submit(sum);   //提交任务
Long r = future.get();                      //获取返回值
fjpool.shutdown();
}
```

上面的代码用到了 RecursiveAction 和 RecursiveTask 两个类，它们都继承自抽象类 ForkJoinTask，用到了其中关键的接口 fork()、join()。二者的区别是一个有返回值，一个没有返回值。代码如下所示。

```
public abstract class RecursiveAction extends ForkJoinTask<Void> {…}
public abstract class RecursiveTask<V> extends ForkJoinTask<V> {…}
```

RecursiveAction/RecursiveTask 类继承体系如图 7-1 所示，其中 AC 表示 Abstract Class。

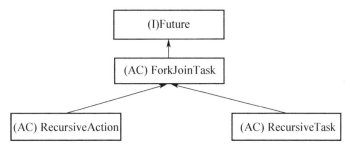

图 7-1　RecursiveAction/RecursiveTask 类继承体系

在 ForkJoinPool 中，对应的接口如下：

```
public <T> ForkJoinTask<T> submit(ForkJoinTask<T> task) { … }
```

7.2　核心数据结构

如图 7-2 所示为 ForkJoinPool 数据结构，不同于 ThreadPoolExecutor，除一个全局的任务队列之外，每个线程还有一个自己的局部队列。

核心数据结构如下所示。

```
public class ForkJoinPool extends AbstractExecutorService {
...
ForkJoinWorkerThread[] workers;                    //线程集合
private ForkJoinTask<?>[] submissionQueue;         //全局队列
volatile int queueBase;                            //队列的尾指针
int queueTop;                                      //队列的头指针
//状态变量，类似于 ThreadPoolExecutor 中的 ctl 变量，后面会详细解释
volatile long ctl;
}
public class ForkJoinWorkerThread extends Thread {
...
ForkJoinTask<?>[] queue;        //每个 Worker 线程有 1 个局部队列
int queueTop;                   //队列的头指针（非 volatile 类型）
volatile int queueBase;         //队列的尾指针（volatile 类型）
final ForkJoinPool pool;        //对所在 ForkJoinPool 的反向引用
final int poolIndex;            //在 Worker[]数组中的 index
}
```

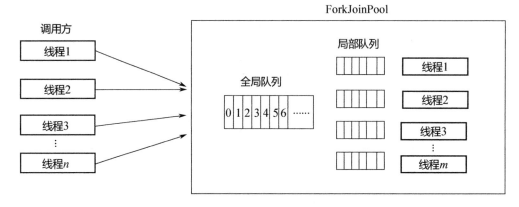

图 7-2 ForkJoinPool 数据结构

下面看一下这些核心数据结构的构造过程。

```
public ForkJoinPool() {
//如果不传并发数，默认等于 CPU 的核数
this(Runtime.getRuntime().availableProcessors(),
    defaultForkJoinWorkerThreadFactory, null, false);
}
public ForkJoinPool(int parallelism) {
    this(parallelism, defaultForkJoinWorkerThreadFactory, null, false);
}
public ForkJoinPool(int parallelism,
```

```
                ForkJoinWorkerThreadFactory factory,
                Thread.UncaughtExceptionHandler handler,
                boolean asyncMode) {
    checkPermission();
    if (factory == null)
        throw new NullPointerException();
    if (parallelism <= 0 || parallelism > MAX_ID)
        throw new IllegalArgumentException();
    this.parallelism = parallelism;
    this.factory = factory;
    this.ueh = handler;
    this.locallyFifo = asyncMode;
    //ctl 变量很关键,后面会详细分析
    long np = (long)(-parallelism);
    this.ctl = ((np << AC_SHIFT) & AC_MASK) | ((np << TC_SHIFT) & TC_MASK);
    //初始全局队列的大小,INITIAL_QUEUE_CAPACITY = 8
    this.submissionQueue = new ForkJoinTask<?>[INITIAL_QUEUE_CAPACITY];
    int n = parallelism << 1;   //n 是 parallelism 的 2 倍
    if (n >= MAX_ID)
        n = MAX_ID;
    else {
        n |= n >>> 1; n |= n >>> 2; n |= n >>> 4; n |= n >>> 8;
    }
    //初始线程池的容量为 2 倍的 parallelism。注意,这个地方并没有创建任何线程,只是一个引
    //用类型的数组而已
    workers = new ForkJoinWorkerThread[n + 1];

    this.submissionLock = new ReentrantLock();
    this.termination = submissionLock.newCondition();
    StringBuilder sb = new StringBuilder("ForkJoinPool-");
    sb.append(poolNumberGenerator.incrementAndGet());
    sb.append("-worker-");
    this.workerNamePrefix = sb.toString();
}
```

7.3 工作窃取队列

关于上面的全局队列,有一个关键点需要说明:它并非使用 BlockingQueue,而是基于一个普通的数组得以实现。

这个队列又名工作窃取队列,为 ForkJoinPool 的工作窃取算法提供服务。在 ForkJoinPool

开篇的注释中，Doug Lea 特别提到了工作窃取队列的实现，其陈述来自如下两篇论文："Dynamic Circular Work-Stealing Deque" by Chase and Lev, SPAA 2005 与"Idempotent work stealing" by Michael, Saraswat, and Vechev,PPoPP 2009。读者可以在网上查阅相应论文。

所谓工作窃取算法，是指一个 Worker 线程在执行完毕自己队列中的任务之后，可以窃取其他线程队列中的任务来执行，从而实现负载均衡，以防有的线程很空闲，有的线程很忙。这个过程要用到工作窃取队列，图 7-3 所示为工作窃取队列示意图。

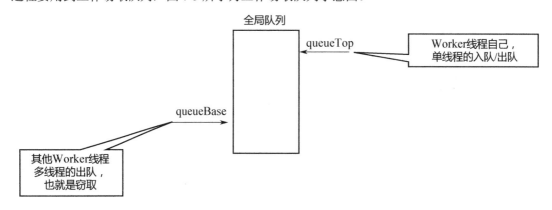

图 7-3　工作窃取队列示意图

这个队列只有三个操作：

（1）Worker 线程自己，在队列头部，通过对 queueTop 指针执行加、减操作，实现入队或出队，这是单线程的。

（2）其他 Worker 线程，在队列尾部，通过对 queueBase 进行累加，实现出队操作，也就是窃取，这是多线程的，需要通过 CAS 操作。

正因为如此，在上面的数据结构定义中，queueTop 不是 volatile 的，queueBase 是 volatile 类型。

这个队列，在 Dynamic Circular Work-Stealing Deque 这篇论文中被称为 dynamic-cyclic-array。之所以这样命名，是因为有两个关键点：

（1）整个队列是环形的，也就是一个数组实现的 RingBuffer。并且 queueBase 会一直累加，不会减小；queueTop 会累加、减小。最后，queueBase、queueTop 的值都会大于整个数组的长度，只是计算数组下标的时候，会取 queueTop & (queue.length–1), queueBase & (queue.length -1)。因为 queue.length 是 2 的整数次方，这里也就是对 queue.length 进行取模操作。

当 queueTop–queueBase = queue.length–1 的时候，队列为满，此时需要扩容；

当 queueTop = queueBase 的时候，队列为空，Worker 线程即将进入阻塞状态。

（2）当队列满了之后会扩容，所以被称为是动态的。但这就涉及一个棘手的问题：多个线程同时在读写这个队列，如何实现在不加锁的情况下一边读写、一边扩容呢？

通过分析工作窃取队列的特性，我们会发现：在 queueBase 一端，是多线程访问的，但它们只会使 queueBase 变大，也就是使队列中的元素变少。所以队列为满，一定发生在 queueTop 一端，对 queueTop 进行累加的时候，这一端却是单线程的！队列的扩容恰好利用了这个单线程的特性！即在扩容过程中，不可能有其他线程对 queueTop 进行修改，只有线程对 queueBase 进行修改！

图 7-4 所示为工作窃取队列扩容示意图。扩容之后，数组长度变成之前的两倍，但 queueTop、queueBase 的值是不变的！通过 queueTop、queueBase 对新的数组长度取模，仍然可以定位到元素在新数组中的位置。

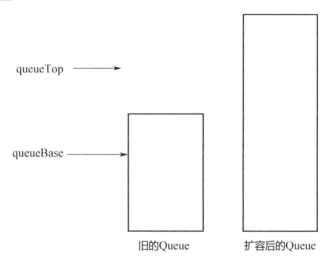

图 7-4　工作窃取队列扩容示意图

下面结合 ForkJoinWorkerThread 扩容的代码进一步分析。

```
private void growQueue() {
ForkJoinTask<?>[] oldQ = queue;
int size = oldQ != null ? oldQ.length << 1 : INITIAL_QUEUE_CAPACITY;
//扩大两倍
if (size > MAXIMUM_QUEUE_CAPACITY)
    throw new RejectedExecutionException("Queue capacity exceeded");
if (size < INITIAL_QUEUE_CAPACITY)
    size = INITIAL_QUEUE_CAPACITY;
ForkJoinTask<?>[] q = queue = new ForkJoinTask<?>[size];
```

```
        int mask = size - 1;
        int top = queueTop;
        int oldMask;
        if (oldQ != null && (oldMask = oldQ.length - 1) >= 0) {
            for (int b = queueBase; b != top; ++b) {           //旧元素逐个复制到新队列
                long u = ((b & oldMask) << ASHIFT) + ABASE;
                Object x = UNSAFE.getObjectVolatile(oldQ, u);  //取旧数组中的元素
                if (x != null && UNSAFE.compareAndSwapObject(oldQ, u, x, null))
                    //旧数组元素置为null
                    UNSAFE.putObjectVolatile
                        (q, ((b & mask) << ASHIFT) + ABASE, x);   //赋值到新数组
            }
        }
    }
}
```

在上面的代码中有两个关键点：

（1）扩容之后的新数组还是空的时候，就已经赋给了成员变量 queue。而 queueTop、queueBase 的值是不变的，这意味着，其他窃取线程若此时来窃取任务，取到的将全是 null，即取不到任务。不过，虽然此时窃取不到，可以阻塞一会儿，待扩容完成就可以窃取到了，不会影响整个算法的正确性。

（2）在把旧数组的元素复制过来之前，先通过 CAS 操作把旧数组中的该元素置为 null。只有 CAS 成功置为 null 了，才能赋值到新数组。这样可以避免同 1 个元素在旧数组、新数组中各有 1 份。1 个窃取线程还在读旧数组，另 1 个窃取线程读取新数组，导致同 1 个元素被 2 个线程重复窃取。

7.4 ForkJoinPool 状态控制

7.4.1 状态变量 ctl 解析

类似于 ThreadPoolExecutor，在 ForkJoinPool 中也有一个 ctl 变量负责表达 ForkJoinPool 的整个生命周期和相关的各种状态。不过 ctl 变量更加复杂，是一个 long 型变量，代码如下所示。

```
public class ForkJoinPool extends AbstractExecutorService {
…
volatile long ctl;
private static final int AC_SHIFT    = 48;
private static final int TC_SHIFT    = 32;
private static final int ST_SHIFT    = 31;
```

第 7 章 ForkJoinPool

```
private static final int  EC_SHIFT   = 16;

private static final int  MAX_ID     = 0x7fff;
private static final int  SMASK      = 0xffff;
private static final int  SHORT_SIGN = 1 << 15;
private static final int  INT_SIGN   = 1 << 31;
private static final long STOP_BIT   = 0x0001L << ST_SHIFT;
private static final long AC_MASK    = ((long)SMASK) << AC_SHIFT;
private static final long TC_MASK    = ((long)SMASK) << TC_SHIFT;

private static final long TC_UNIT    = 1L << TC_SHIFT;
private static final long AC_UNIT    = 1L << AC_SHIFT;

// masks and units for dealing with u = (int)(ctl >>> 32)
private static final int  UAC_SHIFT  = AC_SHIFT - 32;
private static final int  UTC_SHIFT  = TC_SHIFT - 32;
private static final int  UAC_MASK   = SMASK << UAC_SHIFT;
private static final int  UTC_MASK   = SMASK << UTC_SHIFT;
private static final int  UAC_UNIT   = 1 << UAC_SHIFT;
private static final int  UTC_UNIT   = 1 << UTC_SHIFT;

private static final int  E_MASK     = 0x7fffffff;
private static final int  EC_UNIT    = 1 << EC_SHIFT;
    public ForkJoinPool(int parallelism,…) {
…
    long np = (long)(-parallelism);
    this.ctl = ((np << AC_SHIFT) & AC_MASK) | ((np << TC_SHIFT) & TC_MASK);
    }
}
```

图 7-5 所示为 ctl 变量的 64 个比特位含义示意图。这 64 个比特位被分成五部分：

AC：最高的 16 个比特位，表示 Active 线程数–parallelism，parallelism 是上面的构造函数传进去的参数；

TC：次高的 16 个比特位，表示 Total 线程数–parallelism；

ST：1 个比特位，如果是 1，表示整个 ForkJoinPool 正在关闭；

EC：15 个比特位，表示阻塞栈的栈顶线程的 wait count(关于什么是 wait count，接下来的 7.4.2 章节会解释)；

ID：16 个比特位，表示阻塞栈的栈顶线程对应的 poolIndex。

什么叫阻塞栈呢？下面来详细解释：

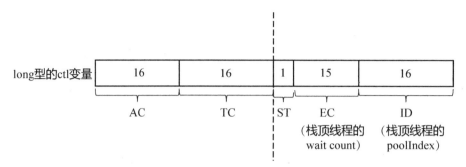

图 7-5　ctl 变量的 64 个比特位含义示意图

7.4.2　阻塞栈–Treiber Stack

要实现多个线程的阻塞、唤醒，除了 park/unpark 这一对操作原语，还需要一个无锁链表实现的阻塞队列，把所有阻塞的线程串在一起。

在 ForkJoinPool 中，没有使用阻塞队列，而是使用了阻塞栈。把所有空闲的 Worker 线程放在一个栈里面，这个栈同样通过链表来实现，名为 Treiber Stack。在 4.5 节讲解 Phaser 的实现原理的时候，也用过这个数据结构。

图 7-6 所示为所有阻塞的 Worker 线程组成的 Treiber Stack。

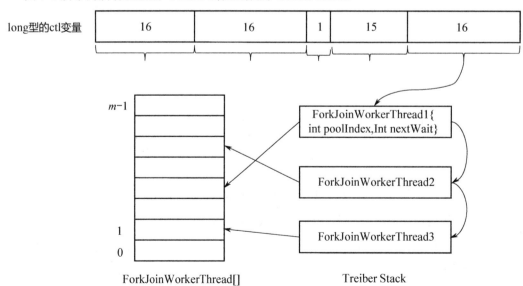

图 7-6　所有阻塞的 Worer 线程组成的 Treiber Stack

首先，ForkJoinWorkerThread 有一个 poolIndex 变量，记录了自己在 ForkJoinWorkerThread[] 数组中的下标位置，poolIndex 变量就相当于每个 ForkJoinPoolWorkerThread 对象的地址；其次，ForkJoinWorkerThread 还有一个 nextWait 变量，记录了前一个阻塞线程的 poolIndex，这个 nextWait 变量就相当于链表的 next 指针，把所有的阻塞线程串联在一起，组成一个 Treiber Stack。

最后，ctl 变量的最低 16 位，记录了栈的栈顶线程的 poolIndex；中间的 15 位，记录了栈顶线程被阻塞的次数，也称为 wait count。

7.4.3 ctl 变量的初始值

在 7.4.1 节的构造函数中，有如下的代码：

```
long np = (long)(-parallelism);
this.ctl = ((np << AC_SHIFT) & AC_MASK) | ((np << TC_SHIFT) & TC_MASK);
```

因为在初始的时候，ForkJoinPool 中的线程个数为 0，所以 AC = 0–parallelism，TC = 0–parallelism。这意味着只有高 32 位的 AC、TC 两个部分填充了值，低 32 位都是 0 填充。

7.4.4 ForkJoinWorkThread 状态与个数分析

在 ThreadPoolExecutor 中，有 corePoolSize 和 maxmiumPoolSize 两个参数联合控制总的线程数，而在 ForkJoinPool 中只传入了一个 parallelism 参数，且这个参数并不是实际的线程数。那么，ForkJoinPool 在实际的运行过程中，线程数究竟是由哪些因素决定的呢？

要回答这个问题，先得明白 ForkJoinPool 中的线程都可能有哪几种状态？可能的状态有三种：

（1）空闲状态（放在 Treiber Stack 里面）。

（2）活跃状态（正在执行某个 ForkJoinTask，未阻塞）。

（3）阻塞状态（正在执行某个 ForkJoinTask，但阻塞了，于是调用 join，等待另外一个任务的结果返回）。

ctl 变量很好地反映出了三种状态：

高 32 位：u = (int)(ctl >>> 32)，然后 u 又拆分成 tc、ac 两个 16 位；

低 32 位：e =(int)ctl。

（1）e > 0，说明 Treiber Stack 不为空，有空闲线程；e = 0，说明没有空闲线程；

（2）ac > 0，说明有活跃线程；ac <= 0，说明没有空闲线程，并且还未超出 parallelism；

（3）tc > 0，说明总线程数 > parallelism。

tc 与 ac 的差值,也就是总线程数与活跃线程数的差异,在 ForkJoinPool 中有另外一个变量 blockedCount 记录,如下:

```
volatile int blockedCount;
```

所以,通过 crl 和 blockedCount 这两个变量,可以知道在整个 ForkJoinPool 中,所有空闲线程、活跃线程以及阻塞线程的数量。

当一个新任务到来时,发现既没有空闲线程,也没有活跃线程,所有线程都阻塞着,在等待任务返回,此时便会开新线程来执行任务。

接下来,我们将结合代码详细了解任务是如何提交的,在执行过程中,线程是如何开启,以及如何执行任务的。在此之前,先了解一下线程的阻塞和唤醒机制。

7.5 Worker 线程的阻塞—唤醒机制

ForkerJoinPool 没有使用 BlockingQueue,也就不曾利用其阻塞—唤醒机制,而是利用了 park/unpark 原语,并自行实现了 Treiber Stack。下面进行详细分析 ForkerJoinPool,在阻塞和唤醒的时候,分别是如何入栈的。

7.5.1 阻塞-入栈

当一个线程窃取不到任何任务,也就是处于空闲状态时就会阻塞,入栈。其核心逻辑在 tryAwaitWork 函数里。

```
private boolean tryAwaitWork(ForkJoinWorkerThread w, long c) {
    int v = w.eventCount;
    w.nextWait = (int)c;   //关键的第一句:使当前线程的 nextWait 指向栈顶
    long nc = (long)(v & E_MASK) | ((c - AC_UNIT) & (AC_MASK|TC_MASK));
    //nc 即 new ctl
    if (ctl != c || !UNSAFE.compareAndSwapLong(this, ctlOffset, c, nc)) {
        long d = ctl; // return true if lost to a deq, to force scan
        return (int)d != (int)c && ((d - c) & AC_MASK) >= 0L;
    }
    for (int sc = w.stealCount; sc != 0;) {      //统计 stealCount
        long s = stealCount;
        if (UNSAFE.compareAndSwapLong(this, stealCountOffset, s, s + sc))
            sc = w.stealCount = 0;
        else if (w.eventCount != v)
            return true;
    }
```

```
    if ((!shutdown || !tryTerminate(false)) &&
        (int)c != 0 && parallelism + (int)(nc >> AC_SHIFT) == 0 &&
        blockedCount == 0 && quiescerCount == 0)
        idleAwaitWork(w, nc, c, v);             //整个ForkJoinPool是否静默
    for (boolean rescanned = false;;) {         //阻塞之前,重新扫了一遍队列
        if (w.eventCount != v)
            return true;
        if (!rescanned) {
            int g = scanGuard, m = g & SMASK;
            ForkJoinWorkerThread[] ws = workers;
            if (ws != null && m < ws.length) {
                rescanned = true;
                for (int i = 0; i <= m; ++i) {
                    ForkJoinWorkerThread u = ws[i];
                    if (u != null) {
                        if (u.queueBase != u.queueTop &&
                            !tryReleaseWaiter())
                            rescanned = false; // contended
                        if (w.eventCount != v)
                            return true;
                    }
                }
            }
            if (scanGuard != g ||
                (queueBase != queueTop && !tryReleaseWaiter()))
                rescanned = false;
            if (!rescanned)
                Thread.yield();
            else
                Thread.interrupted();
        }
        else {
            w.parked = true;
                if (w.eventCount != v) {
                    w.parked = false;
                    return true;
                }
            LockSupport.park(this);     //关键的第二句:进入阻塞
            rescanned = w.parked = false;   //从阻塞中唤醒
        }
    }
}
```

上面的代码之所以复杂，主要是因为它将线程入栈、阻塞之外，还做了很多额外的事情。

函数的第一个参数就是要阻塞的线程，第二个参数是当前的 ctl 变量的值。入栈，也就是 3 步。

第 1 步：w.nextWait = (int)c，即使 w.nextWait 指针，指向栈顶；

第 2 步：long nc = (long)(v & E_MASK) | ((c - AC_UNIT) & (AC_MASK|TC_MASK))，即新的栈顶就是当前的线程 w，同时把 ac 的值减 1，即活跃线程数减 1；

第 3 步：将 nc 通过 CAS 赋值给 ctl 变量 UNSAFE.compareAndSwapLong(this, ctlOffset, c, nc)。

此时入栈成功，ctl 变量更新成功。最后，调用 LockSupport.park(this)阻塞自己。但在这期间，做了很多其他事情：

（1）统计 stealCount。每个 Worker 线程内部都有一个 stealCount 变量，记录该线程窃取了多少个任务。ForkJoinPool 也有一个 stealCount 变量，在线程阻塞之前，把线程的 stealCount 累加到 ForkJoinPool 的 stealCount。这个变量只是一个统计变量，对整体逻辑没有影响。

（2）判断 ForkJoinPool 是否关闭，以及是否所有线程都处于空闲状态，整个 ForkJoinPool 是否处于静默状态。关于这点，后面会再详细讨论。

（3）在阻塞之前，为了保险，又重新扫描了一遍队列，观察是否有任务可以执行。此处涉及很多逻辑，后面会再详细讨论。

7.5.2 唤醒-出栈

在新的任务到来之后，空闲的线程被唤醒，其核心逻辑在 signalWork 函数里面。

```
final void signalWork() {
    long c; int e, u;
    while ((((e = (int)(c = ctl)) | (u = (int)(c >>> 32))) &
        (INT_SIGN|SHORT_SIGN)) == (INT_SIGN|SHORT_SIGN) && e >= 0) {
        if (e > 0) {
            int i; ForkJoinWorkerThread w; ForkJoinWorkerThread[] ws;
            if ((ws = workers) == null ||
                (i = ~e & SMASK) >= ws.length ||   //e的低16位存储的是栈顶元素的poolIndex
                (w = ws[i]) == null)    //取栈顶的元素，如果为空，则直接返回
                break;
            long nc = (((long)(w.nextWait & E_MASK)) |
                       ((long)(u + UAC_UNIT) << 32));
            if (w.eventCount == e &&
                UNSAFE.compareAndSwapLong(this, ctlOffset, c, nc)) {
```

```
            w.eventCount = (e + EC_UNIT) & E_MASK;
            if (w.parked)
                UNSAFE.unpark(w);    //有阻塞的线程，唤醒其中一个
            break;
        }
    }
    else if (UNSAFE.compareAndSwapLong
           (this, ctlOffset, c,
            (long)(((u + UTC_UNIT) & UTC_MASK) |
                ((u + UAC_UNIT) & UAC_MASK)) << 32)) {
        addWorker();      //没有阻塞的线程，开一个新线程
        break;
    }
}
```

该函数的参数为空，它会唤醒栈顶的线程，如上文所述：e > 0，说明栈不为空，此时 e 的最低 16 位，存储的是栈顶线程的 poolIndex，取出来唤醒；e = 0，说明栈为空，此时开一个新线程。

7.6 任务的提交过程分析

在明白了工作窃取队列、ctl 变量的各种状态、Worker 的各种状态，以及线程阻塞—唤醒机制之后，接下来综合这些知识，详细分析任务的提交和执行过程。

关于任务的提交，ForkJoinPool 最外层的接口如下所示。

```
public <T> ForkJoinTask<T> submit(ForkJoinTask<T> task) {
    if (task == null)
        throw new NullPointerException();
    forkOrSubmit(task);
    return task;
}
private <T> void forkOrSubmit(ForkJoinTask<T> task) {
    ForkJoinWorkerThread w;
    Thread t = Thread.currentThread();
    if (shutdown)
        throw new RejectedExecutionException();
    if ((t instanceof ForkJoinWorkerThread) &&         //线程池内部线程
        (w = (ForkJoinWorkerThread)t).pool == this)
        w.pushTask(task);          //加入该线程的局部队列
    else                           //外部调用者线程
```

```
        addSubmission(task);      //加入全局队列
}
```

forkOrSubmit(ForkJoinTask<T> task)中，fork 表示分叉，即是内部生成的任务，submit 是外部提交的任务。

如何区分一个任务是内部任务，还是外部任务呢？可以通过调用该函数的线程类型判断。如果线程类型是 ForkJoinWorkerThread，说明是线程池内部的某个线程在调用该函数，则把该任务放入该线程的局部队列；否则，是外部线程在调用该函数，则将该任务加入全局队列。

7.6.1 内部提交任务 pushTask

内部提交任务，即上面的 w.pushTask(task)，会放入该线程的工作窃取队列中，代码如下所示。

```
public class ForkJoinWorkerThread extends Thread {
...
final void pushTask(ForkJoinTask<?> t) {
ForkJoinTask<?>[] q; int s, m;
if ((q = queue) != null) {
    long u = (((s = queueTop) & (m = q.length - 1)) << ASHIFT) + ABASE;
    UNSAFE.putOrderedObject(q, u, t);
    queueTop = s + 1;  //注意：因为是单线程，pushTask 不需要加锁，直接累加 queueTop
    if ((s -= queueBase) <= 2)
        pool.signalWork();      //queueTop-queueBase <=2,之前是空的，通知空闲线程
    else if (s == m)
        growQueue();           //队列扩容，前面已经分析
    }
}
}
```

signalWork()函数在 7.5 节中已讲，growQueue()函数在 7.3 节中已讲。由于工作窃取队列的特性，其对 queueTop 的操作是单线程的，所以此处不需要执行 CAS 操作。此外，当 queueTop–queueBase = 0 的时候，队列为空，此处为了保险，写作 queueTop–queueBase <= 2，不影响正确性。

7.6.2 外部提交任务 addSubmission

```
private void addSubmission(ForkJoinTask<?> t) {
final ReentrantLock lock = this.submissionLock;
lock.lock();
try {
    ForkJoinTask<?>[] q; int s, m;
    if ((q = submissionQueue) != null) {
```

```
            long u = (((s = queueTop) & (m = q.length-1)) << ASHIFT)+ABASE;
            UNSAFE.putOrderedObject(q, u, t);
            queueTop = s + 1;                    //入队，累加 queueTop
            if (s - queueBase == m)              //队列已满
                growSubmissionQueue();           //扩容
        }
    } finally {
        lock.unlock();
    }
    signalWork();                                //通知空闲线程来取
}
```

外部多个线程会调用该函数，所以要加锁，入队列和扩容的逻辑和线程内部的队列基本相同。最后，调用 signalWork()，通知一个空闲线程来取，这点在 7.5 节已讲。

7.7 工作窃取算法：任务的执行过程分析

全局队列有任务，局部队列也有任务，每一个 Worker 线程都会不间断地扫描这些队列，窃取任务来执行。下面从 Worker 线程的 run 函数开始分析：

```
public class ForkJoinWorkerThread extends Thread {
...
public void run() {
Throwable exception = null;
try {
    onStart();
    pool.work(this);
} catch (Throwable ex) {
    exception = ex;
} finally {
    onTermination(exception);
    }
  }
}
```

run() 函数调用的是所在 ForkJoinPool 的 work 函数，如下所示。

```
public class ForkJoinPool extends AbstractExecutorService {
...
final void work(ForkJoinWorkerThread w) {
    boolean swept = false;            //swept 表示扫描了所有队列，未发现一个任务
long c;
while (!w.terminate && (int)(c = ctl) >= 0) {
```

```
            int a;
            if (!swept && (a = (int)(c >> AC_SHIFT)) <= 0)
                swept = scan(w, a);              //a 就是 ac
            else if (tryAwaitWork(w, c))         //一个任务都没有，进入空闲、阻塞状态
                swept = false;
        }
    }
}
```

只要 w.terminate 不为 true，并且 ctl 变量的低 32 位大于或等于 0，即 shutdown 标志位不为 1，就一直在 while 循环里面，不断扫描任务。如扫描到了，则执行任务；如扫描不到，则调用 tryAwaitWork 进入阻塞状态，阻塞被唤醒之后，继续扫描。

tryAwaitWork 函数在 7.5 节中已讲，下面详细看扫描过程 scan(w, a)。

```
private boolean scan(ForkJoinWorkerThread w, int a) {
    int g = scanGuard;
    int m = (parallelism == 1 - a && blockedCount == 0) ? 0 : g & SMASK;
    ForkJoinWorkerThread[] ws = workers;
    if (ws == null || ws.length <= m)
        return false;
    for (int r = w.seed, k = r, j = -(m + m); j <= m + m; ++j) {  //扫描局部队列
        ForkJoinTask<?> t; ForkJoinTask<?>[] q; int b, i;
        ForkJoinWorkerThread v = ws[k & m];    //注意:m 是离最大下标最近的 2 的整数次方
                                               //-1，当 m 以后到数组最大长度的位置，即
                                               //[m,ws.length -1]都是 null，不需要遍历
        if (v != null && (b = v.queueBase) != v.queueTop &&
            (q = v.queue) != null && (i = (q.length - 1) & b) >= 0) {
            long u = (i << ASHIFT) + ABASE;
            if ((t = q[i]) != null && v.queueBase == b &&
                UNSAFE.compareAndSwapObject(q, u, t, null)) {
                int d = (v.queueBase = b + 1) - v.queueTop;
                v.stealHint = w.poolIndex;
                if (d != 0)
                    signalWork();
                w.execTask(t);      //窃取到一个内部任务，执行，然后返回 false
            }
            r ^= r << 13; r ^= r >>> 17; w.seed = r ^ (r << 5);
            return false;
        }
        else if (j < 0) {
            r ^= r << 13; r ^= r >>> 17; k = r ^= r << 5;
        }
```

```
            else
                ++k;
        }
        if (scanGuard != g)          //扫描队列之后，发现队列已经发生了修改，返回false。之后会
                                     //再次进入scan函数
            return false;
        else {                       //局部队列都是空的，再扫描全局队列
            ForkJoinTask<?> t; ForkJoinTask<?>[] q; int b, i;
            if ((b = queueBase) != queueTop &&
                (q = submissionQueue) != null &&
                (i = (q.length - 1) & b) >= 0) {
                long u = (i << ASHIFT) + ABASE;
                if ((t = q[i]) != null && queueBase == b &&
                    UNSAFE.compareAndSwapObject(q, u, t, null)) {
                    queueBase = b + 1;
                    w.execTask(t);   //窃取到一个外部任务，执行，然后返回false
                }
                return false;
            }
            return true;             //只有当局部队列全为空，全局队列也为空时，才返回true
        }
    }
```

首先，整个函数只有在最后时返回true，在其他时刻都是返回false。返回true，意味着没有任何任务可窃取，回到外面的while循环，进入空闲等待状态；返回false，回到外面的while循环，不会空闲等待，而是再次进入scan函数，继续进行扫描。

其次，在扫描局部队列的时候，不是从0开始扫描，而是随机选取了一个下标作为起点。这是因为如果所有线程都从0扫描到m，会提高冲突概率。但要注意，尽管不是从0开始扫描，但j是从-2m循环到2m的，能够保证在[0,m]之间来回各扫描一遍，不会漏掉某个局部队列。

然后，在整个代码中反复出现一个变量scanGuard。这个变量很关键，但也比较晦涩。要明白这个变量的含义，就需要了解一种锁：顺序锁。

7.7.1 顺序锁 SeqLock

顺序锁有点类似于JDK 8引入的StampedLock，也是乐观读的思想。通过一个sequence number来控制对共享数据的访问，具体来说，就是：

（1）读线程在读取共享数据之前先读取sequence number，在读取数据之后再读一次sequence number，如果两次的值不同，说明在此期间有其他线程修改了数据，此次读取数据无效，重新读取；

（2）写线程，在写入数据之前，累加一次 sequence number，在写入数据之后，再累加一次 sequence number。

最初，sequence number = 0，读线程不会修改 sequence number，而一个写线程会累加两次 sequence number，所以 sequence number 始终是偶数。如果 sequence number 是奇数，说明当前某个写线程正在修改数据，其他写线程被互斥了。

所以，对于写线程而言，发现 sequence number 是奇数，就不能修改共享数据了。对于读线程而言，发现 sequence number 是奇数，也不能再读取数据；如果发现 sequence number 是偶数，那么在读取数据前后分别读取一次 sequence number，如果两次的值相同，则读取成功，否则重新读取。

7.7.2 scanGuard 解析

scanGuard 变量充当了顺序锁的 sequence number 的功能，共享数据就是 Worker 线程数组 ws。所以，在上面的 scan 函数中，在函数开始的时候，读取了一次 scanGuard，扫描完 ws 对应的队列，又读取了一次 scanGuard，发现两次的值不同。说明在这期间 ws 的数组发生了变化，可能是新加了线程，ws 数组扩容了，于是返回 false，重新进入 scan 函数。

当然，scanGuard 变量不仅承担了顺序锁的功能，还表示距离数组中元素最大下标最近的 2 的整数次方-1。具体来说，就是：

（1）其低 16 位，m = scanGuard & SMASK，其值等于距离数组最大下标最近的 2 的整数次方–1。这句话有些拗口，举个例子来说明：假设数组长度是 1024（数组长度始终是 2 的整数次方），里面只装了 10 个线程，最大下标是 12（意味着在 0～12 里面，有 3 个格子是 null 的，这 3 个线程已经死亡，剩下 10 个线程，12 以后的所有格子也都是 null）。那么 m = 16–1 = 15。15 是离 12 最近的，2 的 4 次方–1。

之所以这样记，是为了避免不必要的数组遍历，在这里只需要从 0 遍历到 15 就可以了，而不需要从 0 遍历到 1024；之所以是 2 的整数次方–1，是为了取模优化，这个技巧在整个 JUC 的代码中反复用到，也就是上面代码中 ws[k & m]，等价于 ws[k % (m+1)]。很显然，当数组中装满线程的时候，数组长度等于线程个数，此时 m = 数组长度–1。

（2）第 17 位，也就是 SG_UNIT 位置，标识当前是否有线程在修改 ws 数组。这里使用的其实是 SeqLock 的变体，用第 17 位来控制多个写线程的互斥，用低 16 位来检测在读的过程中是否有写线程修改 ws 数组。

```
public class ForkJoinPool extends AbstractExecutorService {
…
volatile int scanGuard;
```

```
private static final int SG_UNIT = 1 << 16;
   private static final int  SMASK =  0xffff;
}
```

ws 数组的修改主要发生在把一个 Worker 线程加入 ws 数组的时候,下面看一下在这个过程中,scanGuard 变量是如何起作用的。每 1 个 Worker 线程在创建的时候,都会调用下面的函数,把自己注册到 ForkJoinPool 的 ws 数组里面,下面来看具体过程。

```
final int registerWorker(ForkJoinWorkerThread w) {
   for (int g;;) {
    ForkJoinWorkerThread[] ws;
    if (((g = scanGuard) & SG_UNIT) == 0 &&   //第17位为0,尝试置为1
        UNSAFE.compareAndSwapInt(this, scanGuardOffset,
                     g, g | SG_UNIT)) {
        int k = nextWorkerIndex;
        try {
           if ((ws = workers) != null) {
               int n = ws.length;
               if (k < 0 || k >= n || ws[k] != null) {
                   for (k = 0; k < n && ws[k] != null; ++k)
                      ;
                   if (k == n)   //所有位置都不为null,扩容2倍
                      ws = workers = Arrays.copyOf(ws, n << 1);
               }
               ws[k] = w;   //将w注册到第k个位置
               nextWorkerIndex = k + 1; //记录下一个可以注册的位置
               int m = g & SMASK;
               g = (k > m) ? ((m << 1) + 1) & SMASK : g + (SG_UNIT<<1); //新的g
           }
        } finally {
           scanGuard = g;
        }
        return k;
     }
     else if ((ws = workers) != null) { //有其他线程正在修改ws,做些其他额外工作
        for (ForkJoinWorkerThread u : ws) {
           if (u != null && u.queueBase != u.queueTop) {
               if (tryReleaseWaiter())
                   break;
           }
        }
     }
   }
}
```

关于上面的函数，有几点说明：

（1）在写入 ws 之前，先判断 scanGuard 的第 17 位是否为 0。若不为 0，说明有其他线程正在写入 ws；若为 0，则尝试用 CAS，置为 1。

（2）置为 1 成功之后，从下标 0 开始遍历 ws 数组，找到第一个为 null 的位置，把 w 放进去。如果遍历整个数组，未发现有 null 的位置，则对数组进行 2 倍的扩容。

（3）ws 修改成功之后，再恢复 scanGuard 的值。

g = (k > m) ? ((m << 1) + 1) & SMASK : g + (SG_UNIT<<1);

这个地方较难理解，需要详细说明：k > m，说明方才扩容了，此时 m 一定等于旧的数组长度–1，此时 g = ((m << 1) +1) & SMASK。举个例子，数组长度本来是 8，现在扩容到 16，m = 8–1 = 7，g = 7<<1 +1 = 15 = 16–1，同时把第 17 位置回到 0；k <= m，说明没有扩容，此时 g = g + (SG_UNIT << 1)，意为低 16 位没有变化，还是 7，第 17 位置回到 0，同时把第 18 位置为 1 了，表明被某个线程修改过 1 次。

有一点很关键，不论是否扩容，在修改 ws 之前，都需要把第 17 位置为 1，修改之后，第 17 位置回到 0。

7.8　ForkJoinTask 的 fork/join

如果局部队列、全局中的任务全部是相互独立的，就很简单了。但问题是，对于分治算法来说，分解出来的一个个任务并不是独立的，而是相互依赖，一个任务的完成要依赖另一个前置任务的完成。

这种依赖关系是通过 ForkJoinTask 中的 join()来体现的。回到 7.1 节中使用的案例，有这样的代码。

```
protected void compute() {
  if(lo < hi) {
    int pivot = partition(array, lo, hi);   //划分
    SortTask left = new SortTask(array, lo, pivot - 1);
    SortTask right = new SortTask(array, pivot + 1, hi);
    left.fork();
    right.fork();
    left.join();
    right.join();
  }
}
```

线程在执行当前 ForkJoinTask 的时候，产生了 left、right 两个子 Task。所谓 fork，是指把这两个子 Task 放入队列里面，join()则是要等待 2 个子 Task 完成。而子 Task 在执行过程中，会再次产生两个子 Task。如此层层嵌套，类似于递归调用，直到最底层的 Task 计算完成，再一级级返回。

7.8.1 fork

fork()的代码很简单，就是把自己放入当前线程所在的局部队列中。

```
public abstract class ForkJoinTask<V> implements Future<V>, Serializable {
…
public final ForkJoinTask<V> fork() {
((ForkJoinWorkerThread) Thread.currentThread()).pushTask(this);
return this;
}
}
```

7.8.2 join 的层层嵌套

1. join 的层层嵌套阻塞原理

join 会导致线程的层层嵌套阻塞，如图 7-7 所示。

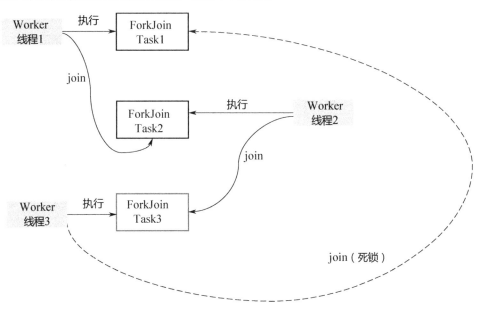

图 7-7　层层嵌套的 join 示意图

线程 1 在执行 ForkJoinTask1，在执行过程中调用了 forkJoinTask2.join()，所以要等 ForkJoinTask2 完成，线程 1 才能返回；

线程 2 在执行 ForkJoinTask2，但由于调用了 forkJoinTask3.join()，只有等 ForkJoinTask3 完成后，线程 2 才能返回；

线程 3 在执行 ForkJoinTask3。

结果是：线程 3 首先执行完，然后线程 2 才能执行完，最后线程 1 再执行完。所有的任务其实组成一个有向无环图 DAG。如果线程 3 调用了 forkJoinTask1.join()，那么会形成环，造成死锁。

那么，这种层次依赖、层次通知的 DAG，在 ForkJoinTask 内部是如何实现的呢？站在 ForkJoinTask 的角度来看，每 1 个 ForkJoinTask，都可能有多个线程在等待它完成，有 1 个线程在执行它。所以每个 ForkJoinTask 就是一个同步对象，线程在调用 join() 的时候，阻塞在这个同步对象上面，执行完成之后，再通过这个同步对象通知所有等待的线程。

如图 7-8 所示，利用 synchronized 关键字和 Java 原生的 wait()/notify 机制，实现了线程的等待-唤醒机制。调用 join() 的这些线程，内部其实是调用 ForkJoinTask 这个对象的 wait()；执行该任务的 Worker 线程，在任务执行完毕之后，顺便调用 notifyAll()。

图 7-8　ForkJoinTask 同步对象示意图

2. ForkJoinTask 的状态解析

要实现 fork()/join() 的这种线程间的同步，对应的 ForkJoinTask 一定是有各种状态的，这个状态变量是实现 fork/join 的基础。

```
public abstract class ForkJoinTask<V> implements Future<V>, Serializable {
...
volatile int status;
private static final int NORMAL      = -1;
private static final int CANCELLED   = -2;
private static final int EXCEPTIONAL = -3;
private static final int SIGNAL      =  1;
}
```

初始时,status = 0。共有五种状态,可以分为两大类:

(1)未完成:0,SIGNAL = 1。即 status >= 0。

0 为初始未完成状态,1 表示有线程阻塞在任务上,等待任务完成。

(2)已完成:NORMAL = -1,CANCELLED = -2,EXCEPTIONAL = -3。即 status < 0。

NORMAL:正常完成;

CANCELLED:任务被取消,完成。

EXCEPTIONAL:任务在执行过程中发生异常,退出也是完成。

所以,通过判断是 status >= 0,还是 status < 0,就可知道任务是否完成,进而决定调用 join() 的线程是否需要被阻塞。

3. join 的详细实现

下面看一下代码的详细实现。

```
public abstract class ForkJoinTask<V> implements Future<V>, Serializable {
...
public final V join() {
if (doJoin() != NORMAL)   //非正常结束,抛出异常
    return reportResult();
else //任务的状态是 NORMAL,正常执行完,取结果
    return getRawResult();
}
```

getRawResult()是 ForkJoinTask 中的一个模板方法,分别被 RecusiveAction 和 RecursiveTask 实现,前者没有返回值,所以返回 null,后者返回一个类型为 V 的 result 变量。

```
public abstract class RecursiveAction extends ForkJoinTask<Void> {
...
public final Void getRawResult() { return null; }
}
public abstract class RecursiveTask<V> extends ForkJoinTask<V> {
...
public final V getRawResult() {
return result;
}
}
```

reportResult() 只是对 getRawResult() 的一个包装,里面多了对 CANCELLED 和 EXCEPTIONAL 这两种异常完成状态的处理。

```java
private V reportResult() {
    int s; Throwable ex;
    if ((s = status) == CANCELLED)
        throw new CancellationException();
    if (s == EXCEPTIONAL && (ex = getThrowableException()) != null)
        UNSAFE.throwException(ex);
    return getRawResult();
}
```

阻塞主要发生在上面的 doJoin() 函数里面。在 dojoin() 里调用 t.join() 的线程会阻塞，然后等待任务 t 执行完成，再唤醒该阻塞线程，doJoin() 返回。

注意：当 doJoin() 返回的时候，就是该任务执行完成的时候，doJoin() 的返回值就是任务的完成状态，也就是上面的 -1、-2、-3 三种状态。

```java
private int doJoin() {
    Thread t; ForkJoinWorkerThread w; int s; boolean completed;
    if ((t = Thread.currentThread()) instanceof ForkJoinWorkerThread) {
        if ((s = status) < 0)          //任务已经完成，不用阻塞了，doJoin()直接返回
            return s;
        if ((w = (ForkJoinWorkerThread)t).unpushTask(this)) {
            try {
                completed = exec();
            } catch (Throwable rex) {
                return setExceptionalCompletion(rex);
            }
            if (completed)
                return setCompletion(NORMAL);
        }
        return w.joinTask(this);       //调用者是一个内部的Worker线程，阻塞在此
    }
    else
        return externalAwaitDone();    //调用者是一个外部线程，阻塞在此
}
```

先看一下 externalAwaitDone()，即外部线程的阻塞过程，相对简单。

```java
private int externalAwaitDone() {
    int s;
    if ((s = status) >= 0) {
        boolean interrupted = false;
        synchronized (this) {
            while ((s = status) >= 0) {
```

```
            if (s == 0)
                UNSAFE.compareAndSwapInt(this, statusOffset,
                                         0, SIGNAL);
            else {
                try {
                    wait();
                } catch (InterruptedException ie) {
                    interrupted = true;
                }
            }
        }
    }
    if (interrupted)
        Thread.currentThread().interrupt();
}
return s;
}
```

可以看到，首先通过 synchronized 关键字对 ForkJoinTask 加锁，之后做了两件事：

（1）把 status 从 0 改成 1(SIGNAL)；

（2）调用 Java 原生的 wait()函数，阻塞该线程。

内部 Worker 线程的阻塞，即上面的 w.joinTask(this)，相比外部线程的阻塞要做更多工作。它的现不在 ForkJoinTask 里面，而是在 ForkJoinWorkerThread 里面。

```
public class ForkJoinWorkerThread extends Thread {
...
final int joinTask(ForkJoinTask<?> joinMe) {
ForkJoinTask<?> prevJoin = currentJoin;
currentJoin = joinMe;
for (int s, retries = MAX_HELP;;) {       //for 死循环，直至 status < 0，才会返回
    if ((s = joinMe.status) < 0) {         //任务完成，返回
        currentJoin = prevJoin;
        return s;
    }
    if (retries > 0) {                     //在阻塞之前，几次自旋重试
        if (queueTop != queueBase) {
            if (!localHelpJoinTask(joinMe))
                retries = 0;
        }
        else if (retries == MAX_HELP >>> 1) {
            --retries;
```

```
            if (tryDeqAndExec(joinMe) >= 0)
                Thread.yield();
        }
        else
            retries = helpJoinTask(joinMe) ? MAX_HELP : retries - 1;
    }
    else {
        retries = MAX_HELP;
        pool.tryAwaitJoin(joinMe);     //阻塞该线程。注意，这里只是尝试，如果不成功，
                                        //会再次回到 for 循环的开始，重新执行
    }
  }
}
```

上面的函数有个关键点：for 里面是死循环，并且只有一个返回点，即只有在 joinMe.status < 0，任务完成之后才可能返回。否则会不断自旋；若自旋之后还不行，就会调用 pool.tryAwaitJoin(joinMe)阻塞。

tryAwaitJoin(joinMe)的代码如下。注意：这里是 tryAwaitJoin(..)，而不是 awaitJoin..)，也就是说，不能保证一定会阻塞成功。如果阻塞不成功，就会返回到 for 循环里，可能再次进入该函数。

```
public class ForkJoinPool extends AbstractExecutorService {
  ...
  final void tryAwaitJoin(ForkJoinTask<?> joinMe) {
    int s;
    Thread.interrupted();
      if (joinMe.status >= 0) {      //只有任务处于未完成状态，才需要执行下面的代码
        if (tryPreBlock()) {          //第 1 部分：阻塞之前的准备工作
          joinMe.tryAwaitDone(0L);    //第 2 部分：阻塞
          postBlock();                //第 3 部分：阻塞唤醒之后的清理工作
        }
        else if ((ctl & STOP_BIT) != 0L)
          joinMe.cancelIgnoringExceptions();
      }
  }
}
```

在上面的函数中，有三个关键部分：阻塞之前、阻塞、阻塞之后。在阻塞之前要做一些准备工作，在阻塞之后要做一些清理工作。具体是哪些工作呢？

```
public class ForkJoinPool extends AbstractExecutorService {
  ...
```

```java
//第1部分：阻塞之前的准备工作。注意，只是尝试，不能保证一定成功。如果不成功，那么会回
//到上面的for循环，然后可能再次进入该函数
private boolean tryPreBlock() {
    int b = blockedCount;
    if (UNSAFE.compareAndSwapInt(this, blockedCountOffset, b, b + 1)) {
        int pc = parallelism;
        do {
            ForkJoinWorkerThread[] ws; ForkJoinWorkerThread w;
            int e, ac, tc, rc, i;
            long c = ctl;
            int u = (int)(c >>> 32);
            if ((e = (int)c) < 0) {    //e < 0,线程池正在关闭，什么都不做，跳出do循环，
                                       //函数会返回false
            }
            else if ((ac = (u >> UAC_SHIFT)) <= 0 && e != 0 &&    //栈中有空闲线程，
                                                                  //唤醒一个线程
                     (ws = workers) != null &&
                     (i = ~e & SMASK) < ws.length &&
                     (w = ws[i]) != null) {
                long nc = ((long)(w.nextWait & E_MASK) |
                           (c & (AC_MASK|TC_MASK)));
                if (w.eventCount == e &&
                    UNSAFE.compareAndSwapLong(this, ctlOffset, c, nc)) {
                    w.eventCount = (e + EC_UNIT) & E_MASK;
                    if (w.parked)
                        UNSAFE.unpark(w);
                    return true;
                }
            }
            else if ((tc = (short)(u >>> UTC_SHIFT)) >= 0 && ac + pc > 1) {
                long nc = ((c - AC_UNIT) & AC_MASK) | (c & ~AC_MASK);
                if (UNSAFE.compareAndSwapLong(this, ctlOffset, c, nc))
                    return true;          //没有空闲线程，但有活跃线程，活跃线程数减1
            }
            else if (tc + pc < MAX_ID) {
                long nc = ((c + TC_UNIT) & TC_MASK) | (c & ~TC_MASK);
                if (UNSAFE.compareAndSwapLong(this, ctlOffset, c, nc)) {
                    addWorker();
                    return true;          //没有空闲线程，也没有活跃线程，要开一个新线程
                }
            }
        } while (!UNSAFE.compareAndSwapInt(this, blockedCountOffset,
```

```
                                b = blockedCount, b - 1));
        }
        return false;
    }

    //第3部分：阻塞唤醒之后的清理工作
    private void postBlock() {
        long c;
        do {} while (!UNSAFE.compareAndSwapLong(this, ctlOffset,
                                    c = ctl, c + AC_UNIT));
        int b;
        do {} while (!UNSAFE.compareAndSwapInt(this, blockedCountOffset,
                                    b = blockedCount, b - 1));
    }
```

上面的 postBlock() 比较简单，只是把活跃线程数加 1，阻塞线程数减 1。

而 tryPreBlock() 相比更复杂：

首先是把阻塞线程数加 1，如果增加失败，整个 tryPreBlock() 就返回 false；如果增加成功，根据线程池的状态 ctl 变量，执行各种对应的操作：

（1）若有空闲线程，从 Treiber Stack 栈顶取出，唤醒。

（2）若无空闲线程，有活跃线程，则只把活跃线程数减 1。这是因为当前的活跃线程马上就要被阻塞了。

（3）如果既无空闲线程，也无活跃线程，意味着所有线程都处于阻塞状态。此时必须开一个新线程，以应对后续的任务。

在 tryPreBlock() 和 postBlock() 之间，就是实际执行阻塞的地方。上文已讲，外部线程的阻塞是通过调用 ForkJoinTask 的 externalAwaitDone() 完成的；内部线程的阻塞调用了 tryAwaitDone(..)，代码如下。两个函数的实现基本类似，都是做了两件事：首先把 status 从 0 改到 1；其次，调用 Java 原生的 wait() 函数，阻塞该线程。读者可以翻阅前面的部分对两个函数进行比较。

```
public abstract class ForkJoinTask<V> implements Future<V>, Serializable {
    ...
    //第2部分：阻塞的核心实现。此处也是tryAwaitDone()，而非AwaitDone()，无法保证一定
    //成功。如果不成功，就会回到for循环，然后可能再次进入该函数
    final void tryAwaitDone(long millis) {
        int s;
        try {
            if (((s = status) > 0 ||
```

```
            (s == 0 &&
             UNSAFE.compareAndSwapInt(this, statusOffset, 0, SIGNAL))) &&
            status > 0) {
            synchronized (this) {
                if (status > 0)    //小于0，任务完成，无须阻塞，返回
                    wait(millis);
            }
        }
    } catch (InterruptedException ie) {
        // caller must check termination
    }
  }
}
```

4. join 的唤醒

调用 t.join() 之后，线程会被阻塞。接下来看另外一个线程在任务 t 执行完毕后如何唤醒阻塞的线程。

```
public abstract class ForkJoinTask<V> implements Future<V>, Serializable {
...
final void doExec() {
    if (status >= 0) {
        boolean completed;
        try {
            completed = exec();
        } catch (Throwable rex) {
            setExceptionalCompletion(rex);
            return;
        }
        if (completed)
            setCompletion(NORMAL);  //任务执行完毕，更新 status，同时通知其他阻塞的线程
    }
}

private int setCompletion(int completion) {
    for (int s;;) {
        if ((s = status) < 0)    //本来就已经完成，什么都不做，直接返回
            return s;
        if (UNSAFE.compareAndSwapInt(this, statusOffset, s, completion)) {
            if (s != 0)
                synchronized (this) { notifyAll(); }   //通知所有被阻塞的线程
            return completion;
```

```
            }
        }
    }
}
```

任务的执行发生在 doExec()函数里面，任务执行完成后，调用一个 setCompletion(..)通知所有等待的线程。这里也做了两件事：

（1）把 status 置为完成状态。也就是 NORMAL、CANCELLED 或者 EXCEPTIONAL。

（2）如果 s != 0，即 s = SIGNAL，说明有线程正在等待这个任务执行完成。调用 Java 原生的 notifyAll()通知所有线程。如果 s = 0，说明没有线程等待这个任务，不需要通知。

7.9 ForkJoinPool 的优雅关闭

同 ThreadPoolExecutor 一样，ForkJoinPool 的关闭也不可能是"瞬时的"，而是需要一个平滑的过渡过程。

7.9.1 关键的 terminate 变量

对于一个 Worker 线程来说，它会在一个 while 循环里面不断轮询队列中的任务，如果有任务，那么执行，处在活跃状态；如果没有任务，则进入空闲等待状态。那么，这个线程如何退出呢？答案在下面的代码中。

```
final void work(ForkJoinWorkerThread w) {
    boolean swept = false;
    long c;
    while (!w.terminate && (int)(c = ctl) >= 0) {
        int a;
        if (!swept && (a = (int)(c >> AC_SHIFT)) <= 0)
            swept = scan(w, a);
        else if (tryAwaitWork(w, c))
            swept = false;
    }
}
```

(int)(c = ctl) < 0，即低 32 位的最高位为 1（参考前面 ctl 变量的解析），说明线程池已经进入了关闭状态。但线程池进入关闭状态，不代表所有的线程都会立马关闭。

为此，在 ForkJoinWorkerThread 里还有一个 terminate 变量，初始为 false。当线程池要关闭的时候，会把相关线程的 terminate 变量置为 true。这样，这些线程就会退出上面的 while 循环，也就会自动退出。

当线程池关闭的时候，什么样的线程可以立马关闭，什么样的线程不能立马关闭呢？这就涉及下面两个函数的区别。

7.9.2　shutdown()与shutdownNow()的区别

```
public void shutdown() {
    checkPermission();
    shutdown = true;
    tryTerminate(false);
}
public List<Runnable> shutdownNow() {
    checkPermission();
    shutdown = true;
    tryTerminate(true);
    return Collections.emptyList();
}
```

二者的代码基本相同，都是调用 tryTerminate(boolean)函数，其中一个传入的是 false，另一个传入的是 true。tryTerminate 意为试图关闭 ForkJoinPool，并不保证一定可以关闭成功。

```
private boolean tryTerminate(boolean now) {
    long c;
    while (((c = ctl) & STOP_BIT) == 0) {
        if (!now) {
            if ((int)(c >> AC_SHIFT) != -parallelism)
                return false;
            if (!shutdown || blockedCount != 0 || quiescerCount != 0 ||
                queueBase != queueTop) {
                if (ctl == c)
                    return false;
                continue;
            }
        }
        if (UNSAFE.compareAndSwapLong(this, ctlOffset, c, c | STOP_BIT))
            startTerminating();
    }

    if ((short)(c >>> TC_SHIFT) == -parallelism) {
        final ReentrantLock lock = this.submissionLock;
        lock.lock();
        try {
            termination.signalAll();      //通知阻塞在awaitTermination(..)的线程，其
                                          //逻辑和前面讲的ThreadPoolExecutor几乎一样
```

```
        } finally {
            lock.unlock();
        }
    }
    return true;
}
```

同 ThreadPoolExecutor 一样,ForkJoinPool 中也有 awaitTermination(…)函数,代码几乎相同,上面函数的最后一段,就是在整个线程池都已关闭,即没有任何线程存活的情况下,通知阻塞在 awaitTermination(..)的外部线程。

如果 now = true,就会执行以下代码。

```
if (UNSAFE.compareAndSwapLong(this, ctlOffset, c, c | STOP_BIT))
    startTerminating();
```

把 ctl 变量低 32 位的最高位,通过 CAS 操作置成 1,然后调用 startTerminating()。

```
private void startTerminating() {
    cancelSubmissions();    //取消全局队列中的所有任务
    for (int pass = 0; pass < 3; ++pass) {
        ForkJoinWorkerThread[] ws = workers;
        if (ws != null) {
            for (ForkJoinWorkerThread w : ws) {
                if (w != null) {
                    w.terminate = true;    //把所有线程的 terminate 置为 true
                    if (pass > 0) {
                        w.cancelTasks();    //线程内部的局部队列中的所有任务也被取消
                        if (pass > 1 && !w.isInterrupted()) {
                            try {
                                w.interrupt();
                            } catch (SecurityException ignore) {
                            }
                        }
                    }
                }
            }
        }
    }
    terminateWaiters();    //唤醒所有等待的空闲线程,这些空闲线程会自动退出
}
```

通过上面代码可以看出,在 startTerminating()中,把全局队列、每个线程的局部队列中的任务都取消了,同时把所有线程的 terminate 置为了 true,唤醒了阻塞栈中所有等待的空闲线程(这

些线程的 terminate 置为了 true，会自动退出）。

如果 now 为 false，tryTerminate(…)会返回 false。只是在最外面的函数 shutdown()里面，把 shutdown 置为了 true。这样，新的任务提交会被拒绝，但现有的任务都会正常执行完成。

所以，最后总结一下：shutdown()只拒绝新提交的任务；shutdownNow()会取消现有的全局队列和局部队列中的任务，同时唤醒所有空闲的线程，让这些线程自动退出。

第 8 章 CompletableFuture

从 JDK 8 开始，在 Concurrent 包中提供了一个强大的异步编程工具 CompletableFuture。在 JDK8 之前，异步编程可以通过线程池和 Future 来实现，但功能还不够强大。CompletableFuture 的出现，使 Java 的异步编程能力向前迈进了一大步。

在探讨 CompletableFuture 的原理之前，先详细看一下 CompletableFuture 的用法，从这些用法中，可以看到相较之前的 Future 有哪些能力得到了提升。

8.1　CompletableFuture 用法

8.1.1　最简单的用法

CompletableFuture 实现了 Future 接口，所以它也具有 Future 的特性：调用 get()方法会阻塞在那，直到结果返回。

```
CompletableFuture<String> completableFuture = new CompletableFuture<String>();
String result = completableFuture.get();   //调用者阻塞，等待结果返回
```

另外 1 个线程调用 complete 方法完成该 Future，则所有阻塞在 get()方法的线程都将获得返回结果。

```
completableFuture.complete("this is a test result")
```

8.1.2　提交任务：runAsync 与 supplyAsync

上面的例子是一个空的任务，下面尝试提交一个真的任务，然后等待结果返回。

例1: runAsync(Runnable)

```
CompletableFuture<Void> future1 = CompletableFuture.runAsync(() -> {
    try {
        System.out.println("test task is running");
        TimeUnit.SECONDS.sleep(1);
    } catch (InterruptedException e) {
        throw new IllegalStateException(e);
    }
});
future1.get();    //主线程阻塞，等待任务执行完成
```

CompletableFuture.runAsync(..)传入的是一个 Runnable 接口，在上面的代码中是使用了 Java 8 的 lambda 表达式的写法，和定义一个 Runnable 对象是等价的。

例2: supplyAsync(Supplier)

```
CompletableFuture<String> future2 = CompletableFuture.supplyAsync(new 
Supplier<String>() {
    public String get() {
        try {
            TimeUnit.SECONDS.sleep(1);
        } catch (InterruptedException e) {
            throw new IllegalStateException(e);
        }
        return "test result";
    }
});
String result = future2.get();    //主线程阻塞，等待结果返回
System.out.println(result);
```

例 2 和例 1 的区别在于，例 2 的任务有返回值。没有返回值的任务，提交的是 Runnable，返回的是 CompletableFuture<Void>；有返回值的任务，提交的是 Supplier，返回的是 CompletableFuture<String>。Supplier 和前面的 Callable 很相似。

通过上面两个例子可以看出，在基本的用法上，CompletableFuture 和 Future 很相似，都可以提交两类任务：一类是无返回值的，另一类是有返回值的。

8.1.3　链式的 CompletableFuture: thenRun、thenAccept 和 thenApply

对于 Future，在提交任务之后，只能调用 get()等结果返回；但对于 CompletableFuture，可以在结果上面再加一个 callback，当得到结果之后，再接着执行 callback。

例1: thenRun(Runnable)

```
CompletableFuture<Void> testFuture = CompletableFuture.supplyAsync(() -> {
```

```java
    try {
        TimeUnit.SECONDS.sleep(1);
    } catch (InterruptedException e) {
        throw new IllegalStateException(e);
    }
    return "test result";
}).thenRun(() -> {
    System.out.println("task finished");
});
```

例2：thenAccept(Consumer)

```java
CompletableFuture<Void> testFuture = CompletableFuture.supplyAsync(() -> {
    try {
        TimeUnit.SECONDS.sleep(1);
    } catch (InterruptedException e) {
        throw new IllegalStateException(e);
    }
    return "test result";
}).thenAccept(result -> {
    System.out.println(result);
});
```

例3：thenApply(Function)

```java
CompletableFuture<String> testFuture = CompletableFuture.supplyAsync(() -> {
    try {
        TimeUnit.SECONDS.sleep(1);
    } catch (InterruptedException e) {
        throw new IllegalStateException(e);
    }
    return "test result";
}).thenApply(result -> {
    return result + "after thenApply";
});
```

三个例子都是在任务执行完成之后，再紧急执行一个 callback，只是 callback 的形式有所区别：

（1）thenRun 后面跟的是一个无参数、无返回值的方法，即 Runnable，所以最终的返回值是 CompletableFuture<Void>类型。

（2）thenAccept 后面跟的是一个有参数、无返回值的方法，称为 Consumer，返回值也是 CompletableFuture<Void>类型。顾名思义，只进不出，所以称为 Consumer；前面的 Supplier，是无参数，有返回值，只出不进，和 Consumer 刚好相反。

（3）thenApply 后面跟的是一个有参数、有返回值的方法，称为 Function。返回值是 CompletableFuture<String>类型。

而参数接收的是前一个任务，即 supplyAsync(..)这个任务的返回值。因此这里只能用 supplyAsync，不能用 runAsync。因为 runAsync 没有返回值，不能为下一个链式方法传入参数。

8.1.4　CompletableFuture 的组合：thenCompose 与 thenCombine

例 1：thenCompose

在上面的例子中，thenApply 接收的是一个 Function，但是这个 Function 的返回值是一个通常的基本数据类型或一个对象，而不是另外一个 CompletableFuture。如果 Function 的返回值也是一个 CompletableFuture，就会出现嵌套的 CompletableFuture。考虑下面的例子：

函数 1：输入 userId，返回 user 对象。但由于是异步调用，所以返回了一个 CompletableFuture<User>对象。

```
CompletableFuture<User> getUsersDetail(String userId) {
    return CompletableFuture.supplyAsync(() -> {
        UserService.getUserDetails(userId);
    });
}
```

函数 2：输入 user 对象，获取该 user 的信用卡积分（Double 类型）。由于是异步调用，返回的不是 Double，而是 CompletableFuture<Double>。

```
CompletableFuture<Double> getCreditRating(User user) {
    return CompletableFuture.supplyAsync(() -> {
        CreditRatingService.getCreditRating(user);
    });
}
```

这两个函数链式调用，代码如下所示。

```
CompletableFuture<CompletableFuture<Double>> result = getUserDetail(userId)
.thenApply(user -> getCreditRating(user))
```

因为函数 getCreditRating(user)返回的是一个 CompletableFuture<Double>类型，所以在链式调用之后，返回值变成一个嵌套的数据类型 CompletableFuture<CompletableFuture<Double>>。

如果希望返回值是一个展平的 CompletableFuture，可以使用 thenCompose，代码如下所示。

```
CompletableFuture<Double> result = getUserDetail(userId)
.thenCompose(user -> getCreditRating(user));
```

下面是 thenCompose 函数的接口定义。

```
public <U> CompletionStage<U> thenCompose
(Function<? super T, ? extends CompletionStage<U>> fn);
```

从该函数的定义可以看出，它传入的参数是一个 Function 类型，并且 Function 的返回值必须是 CompletionStage 的子类，也就是 CompletableFuture 类型。

例 2：thenCombine

thenCombine 函数的接口定义如下，从传入的参数可以看出，它不同于 thenCompose。

```
public <U,V> CompletionStage<V> thenCombine
(CompletionStage<? extends U> other,
 BiFunction<? super T,? super U,? extends V> fn);
```

第 1 个参数是一个 CompletableFuture 类型，第 2 个参数是一个函数，并且是一个 BiFunction，也就是该函数有 2 个输入参数，1 个返回值。

从该接口的定义可以大致推测，它是要在 2 个 CompletableFuture 完成之后，把 2 个 CompletableFuture 的返回值传进去，再额外做一些事情。实例如下：

函数 1：获取体重。

```
CompletableFuture<Double> weightFuture = CompletableFuture.supplyAsync(() -> {
    try {
        TimeUnit.SECONDS.sleep(1);
    } catch (InterruptedException e) {
      throw new IllegalStateException(e);
    }
    return 65.0;
});
```

函数 2：获取身高。

```
CompletableFuture<Double> heightFuture = CompletableFuture.supplyAsync(() -> {
    try {
        TimeUnit.SECONDS.sleep(1);
    } catch (InterruptedException e) {
      throw new IllegalStateException(e);
    }
    return 177.8;
});
```

函数 3：结合身高、体重，计算 BMI 指数。

```
CompletableFuture<Double> bmiFuture = weightFuture
      .thenCombine(heightFuture, (weight, height) -> {
    Double heightInMeter = height/100;
    return weight/(heightInMeter*heightInMeter);
```

        });

        System.out.println("Your BMI is - " + bmiFuture.get());

8.1.5 任意个 CompletableFuture 的组合

上面的 thenCompose 和 thenCombine 只能组合 2 个 CompletableFuture，而接下来的 allOf 和 anyOf 可以组合任意多个 CompletableFuture。函数接口定义如下所示。

```
public static CompletableFuture<Void> allOf(CompletableFuture<?>… cfs) {
   …
}
public static CompletableFuture<Object> anyOf(CompletableFuture<?>… cfs) {
   …
}
```

首先，这两个函数都是静态函数，参数是变长的 CompletableFuture 的集合。其次，allOf 和 anyOf 的区别，前者是"与"，后者是"或"。

例 1：allOf

allOf 的返回值是 CompletableFuture<Void> 类型，这是因为每个传入的 CompletableFuture 的返回值都可能不同，所以组合的结果是无法用某种类型来表示的，索性返回 Void 类型。那么，如何获取每个 CompletableFuture 的执行结果呢？参看下面的例子：

并行地下载 100 个网页。待下载完成之后，统计在 100 个网页中，含有某个单词的网页个数。整个过程可以分为三步：

第 1 步：并行下载 100 个网页。

```
List<String> webPageLinks = Arrays.asList(…)    //url 列表
List<CompletableFuture<String>> pageContentFutures = webPageLinks.stream()
      .map(webPageLink -> downloadWebPage(webPageLink))
      .collect(Collectors.toList());
```

第 2 步：通过 allOf，等待所有网页下载完毕，收集返回结果。

```
CompletableFuture<Void> allFutures = CompletableFuture.allOf(
      pageContentFutures.toArray(new
CompletableFuture[pageContentFutures.size()])
);
```

关键点：因为 allOf 没有返回值，所以通过 thenApply，给 allFutures 附上一个回调函数。在回调函数里面，依次调用每个 future 的 get() 函数，获取到 100 个结果，存入 List<String>。

```
CompletableFuture<List<String>> allPageContentsFuture = allFutures.thenApply
(v -> {
```

```
    return pageContentFutures.stream()
            .map(pageContentFuture -> pageContentFuture.get())
            .collect(Collectors.toList());
});
```

第 3 步：统计在这 100 个网页中，含有单词 "XXX" 的网页的个数。

```
CompletableFuture<Long> countFuture = allPageContentsFuture.thenApply
(pageContents -> {
    return pageContents.stream()
            .filter(pageContent -> pageContent.contains("XXX"))
            .count();
});
System.out.println(countFuture.get());
```

例 2：anyOf

anyOf 的含义是只要有任意一个 CompletableFuture 结束，就可以做接下来的事情，而无须像 AllOf 那样，等待所有的 CompletableFuture 结束。

但由于每个 CompletableFuture 的返回值类型都可能不同，任意一个，意味着无法判断是什么类型，所以 anyOf 的返回值是 CompletableFuture<Object> 类型。

考虑下面的例子。

```
CompletableFuture<String> future1 = CompletableFuture.supplyAsync(() -> {
    try {
        TimeUnit.SECONDS.sleep(2);
    } catch (InterruptedException e) {
      throw new IllegalStateException(e);
    }
    return "Result of Future 1";
});
CompletableFuture<String> future2 = CompletableFuture.supplyAsync(() -> {
    try {
        TimeUnit.SECONDS.sleep(1);
    } catch (InterruptedException e) {
      throw new IllegalStateException(e);
    }
    return "Result of Future 2";
});
CompletableFuture<String> future3 = CompletableFuture.supplyAsync(() -> {
    try {
        TimeUnit.SECONDS.sleep(3);
    } catch (InterruptedException e) {
```

```
        throw new IllegalStateException(e);
    }
    return "Result of Future 3";
});
CompletableFuture<Object> anyOfFuture = CompletableFuture.anyOf(future1,
future2, future3);
System.out.println(anyOfFuture.get());
```

在该例子中，因为 future1、future2、future3 的返回值都是 CompletableFuture<String>类型，所以 anyOf 返回的 Object，一定也是 String 类型。

并且在 3 个 future 中，future2 睡眠时间最短，会最先执行完成，anyOfFuture.get()获取的也就是 future2 的内容。future1、future3 的返回结果被丢弃了。

8.2 四种任务原型

通过上面的例子可以总结出，提交给 CompletableFuture 执行的任务有四种类型：Runnable、Consumer、Supplier、Function。表格 8-1 总结了这四种任务原型的对比。

表 8-1 四种任务原型的对比

四种任务原型	无参数	有参数
无返回值	Runnable 接口 对应的提交方法：runAsync，thenRun	Consumer 接口 对应的提交方法：thenAccept
有返回值	Supplier 接口 对应的提交方法：supplierAsync	Function 接口 对应的提交方法：thenApply

runAsync 与 supplierAsync 是 CompletableFutre 的静态方法；而 thenAccept、thenAsync、thenApply 是 CompletableFutre 的成员方法。

因为初始的时候没有 CompletableFuture 对象，也没有参数可传，所以提交的只能是 Runnable 或者 Supplier，只能是静态方法；

通过静态方法生成 CompletableFuture 对象之后，便可以链式地提交其他任务了，这个时候就可以提交 Runnable、Consumer、Function，且都是成员方法。

8.3 CompletionStage 接口

CompletableFuture 不仅实现了 Future 接口，还实现了 CompletableStage 接口，如下所示。

```java
public class CompletableFuture<T> implements Future<T>, CompletionStage<T> {
…
}
```

CompletionStage 接口定义的正是前面的各种链式方法、组合方法，如下所示。

```java
public interface CompletionStage<T> {
public CompletionStage<Void> thenRun(Runnable action);
public CompletionStage<Void> thenAccept(Consumer<? super T> action);
public <U> CompletionStage<U> thenApply(Function<? super T,? extends U> fn);
//Function 的输入参数必须是? super T 类型，返回值是? Extends U 类型
public <U> CompletionStage<U> thenCompose
(Function<? super T, ? extends CompletionStage<U>> fn);
public <U,V> CompletionStage<V> thenCombine
(CompletionStage<? extends U> other,
 BiFunction<? super T,? super U,? extends V> fn);
…
}
```

关于 CompletionStage 接口，有几个关键点要说明：

（1）所有方法的返回值都是 CompletionStage 类型，也就是它自己。正因为如此，才能实现如下的链式调用：future1.thenApply(xxx).thenApply(xxx).thenCompose(…).thenRun(..)。

（2）thenApply 接受的是一个有输入参数、返回值的 Function。这个 Function 的输入参数，必须是? Super T 类型，也就是 T 或者 T 的父类型，而 T 必须是调用 thenApplycompletableFuture 对象的类型；返回值则必须是? Extends U 类型，也就是 U 或者 U 的子类型，而 U 恰好是 thenApply 的返回值的 CompletionStage 对应的类型。

其他函数，诸如 thenCompose、thenCombine 也是类似的原理。

8.4 CompletableFuture 内部原理

8.4.1 CompletableFuture 的构造：ForkJoinPool

CompletableFuture 中任务的执行同样依靠 ForkJoinPool，代码如下所示。

```java
public class CompletableFuture<T> implements Future<T>, CompletionStage<T> {
private static final Executor asyncPool = useCommonPool ?
ForkJoinPool.commonPool() : new ThreadPerTaskExecutor();
public static <U> CompletableFuture<U> supplyAsync(Supplier<U> supplier) {
return asyncSupplyStage(asyncPool, supplier);
```

```
}
    static <U> CompletableFuture<U> asyncSupplyStage(Executor e, Supplier<U> f) {
        if (f == null) throw new NullPointerException();
        CompletableFuture<U> d = new CompletableFuture<U>();
        e.execute(new AsyncSupply<U>(d, f));   //Supplier 转换为 ForkJoinTask
        return d;
    }
    ...
}
```

通过上面的代码可以看到，asyncPool 是一个 static 类型，supplierAsync、asyncSupplyStage 也都是 static 函数。Static 函数会返回一个 CompletableFuture 类型对象，之后就可以链式调用，CompletionStage 里面的各个方法。

8.4.2 任务类型的适配

ForkJoinPool 接受的任务是 ForkJoinTask 类型，而我们向 CompletableFuture 提交的任务是 Runnable/Supplier/Consumer/Function。因此，肯定需要一个适配机制，把这四种类型的任务转换成 ForkJoinTask，然后提交给 ForkJoinPool，如图 8-1 所示。

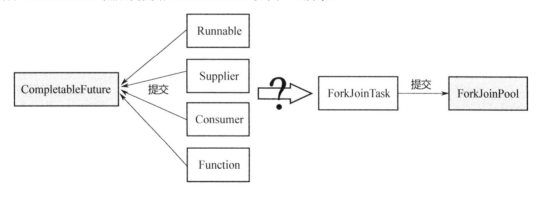

图 8-1　Runnable/Supplier/Consumer/Function 到 ForkJoinTask 的转换问题

为了完成这种转换，在 CompletableFuture 内部定义了一系列的内部类，图 8-2 所示为 CompletableFuture 的各种内部类的继承体系。

在 supplierAsync(..)函数内部，会把一个 Supplier 转换成一个 AsyncSupply，然后提交给 ForkJoinPool 执行；

在 runAsync(..)函数内部,会把一个 Runnable 转换成一个 AsyncRun,然后提交给 ForkJoinPool 执行；

在 thenRun/thenAccept/thenApply 内部，会分别把 Runnable/Consumer/Function 转换成 UniRun/UniAccept/UniApply 对象，然后提交给 ForkJoinPool 执行；

除此之外，还有两种 CompletableFuture 组合的情况，分为"与"和"或"，所以有对应的 Bi 和 Or 类型的 Completion 类型。

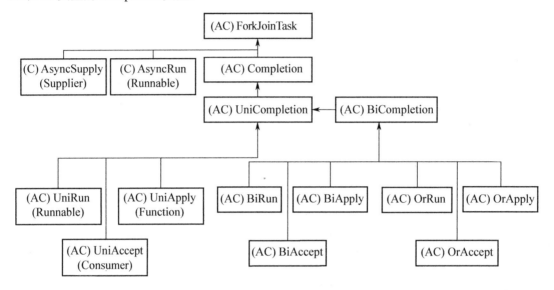

图 8-2　CompletableFuture 的各种内部类的继承体系

下面的代码分别为 UniRun、UniApply、UniAccept 的定义，可以看到，其内部分别封装了 Runnable、Function、Consumer。

```
static final class UniRun<T> extends UniCompletion<T,Void> {
Runnable fn;
...
}
static final class UniApply<T,V> extends UniCompletion<T,V> {
Function<? super T,? extends V> fn;
...
}
static final class UniAccept<T> extends UniCompletion<T,Void> {
Consumer<? super T> fn;
...
}
```

图 8-3 所示为 CompletableFuture 的接口层面和内部实现层面对比。

第 8 章 CompletableFuture

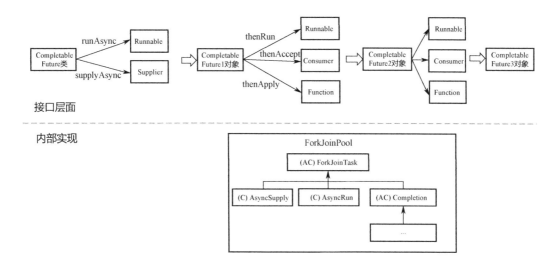

图 8-3 CompletableFuture 的接口层面和内部实现层面对比

8.4.3 任务的链式执行过程分析

下面以 CompletableFuture.supplyAsync(…).thenApply(…).thenRun(…) 链式代码为例，分析整个执行过程。

第 1 步：CompletableFuture future1 = CompletableFuture.supplyAsync(…)

```
public static <U> CompletableFuture<U> supplyAsync(Supplier<U> supplier) {
    return asyncSupplyStage(asyncPool, supplier);
}
static <U> CompletableFuture<U> asyncSupplyStage(Executor e,
                                    Supplier<U> f) {
    if (f == null) throw new NullPointerException();
    CompletableFuture<U> d = new CompletableFuture<U>();
    e.execute(new AsyncSupply<U>(d, f));
    return d;
}
```

在上面的代码中，关键是构造了一个 AsyncSupply 对象，该对象有三个关键点：

（1）它继承自 ForkJoinTask，所以能够提交 ForkJoinPool 来执行。

（2）它封装了 Supplier f，即它所执行任务的具体内容。

（3）该任务的返回值，即 CompletableFuture d，也被封装在里面。

图 8-4 所示为这几个概念之间的关系。ForkJoinPool 执行一个 ForkJoinTask 类型的任务，即 AsyncSupply。该任务的输入就是 Supply，输出结果存放在 CompletableFuture 中。

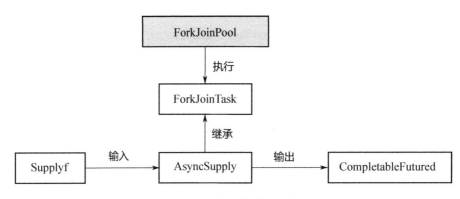

图 8-4　几个概念之间的关系

第 2 步：CompletableFuture future2 = future1.thenApply(…)

第 1 步的返回值，也就是上面代码中的 CompletableFuture d，紧接着调用其成员函数 thenApply。

```
public <U> CompletableFuture<U> thenApply(
Function<? super T,? extends U> fn) {
return uniApplyStage(null, fn);
}
private <V> CompletableFuture<V> uniApplyStage(
Executor e, Function<? super T,? extends V> f) {
if (f == null) throw new NullPointerException();
CompletableFuture<V> d =  new CompletableFuture<V>();
if (e != null || !d.uniApply(this, f, null)) {
    UniApply<T,V> c = new UniApply<T,V>(e, d, this, f);
    push(c);   //把第 2 个任务压入第 1 个任务的执行结果所在的栈
    c.tryFire(SYNC);
}
return d;
}
```

我们知道，必须等第 1 步的任务执行完毕，第 2 步的任务才可以执行。因此，这里提交的任务不可能立即执行，在此处构建了一个 UniApply 对象，也就是一个 ForkJoinTask 类型的任务，这个任务放入了第 1 个任务的栈当中。

```
final void push(UniCompletion<?,?> c) {
if (c != null) {
    while (result == null && !tryPushStack(c))
        lazySetNext(c, null); // clear on failure
}
}
```

第 8 章 CompletableFuture

每一个 CompletableFuture 对象内部都有一个栈，存储着是后续依赖它的任务，如下面代码所示。这个栈也就是 Treiber Stack，这里的 stack 存储的就是栈顶指针。

```
public class CompletableFuture<T> implements Future<T>, CompletionStage<T>
{
...
volatile Completion stack;
...
}
```

上面的 UniApply 对象类似于第 1 步里面的 AsyncSupply，它的构造函数传入了 4 个参数：

第 1 个参数是执行它的 ForkJoinPool；

第 2 个参数是输出一个 CompletableFuture 对象。这个参数，也是 thenApply 函数的返回值，用来链式执行下一个任务；

第 3 个参数是其依赖的前置任务，也就是第 1 步里面提交的任务；

第 4 个参数是输入（也就是一个 Function 对象）。

```
UniApply(Executor executor, CompletableFuture<V> dep,
    CompletableFuture<T> src,
    Function<? super T,? extends V> fn) {
super(executor, dep, src); this.fn = fn;
}
```

UniApply 对象被放入了第 1 步的 CompletableFuture 的栈中，在第 1 步的任务执行完成之后，就会从栈中弹出并执行。下面看一下代码：

```
static final class AsyncSupply<T> extends ForkJoinTask<Void>
    implements Runnable, AsynchronousCompletionTask {
CompletableFuture<T> dep; Supplier<T> fn;
AsyncSupply(CompletableFuture<T> dep, Supplier<T> fn) {
    this.dep = dep; this.fn = fn;
}
public void run() {
CompletableFuture<T> d; Supplier<T> f;
if ((d = dep) != null && (f = fn) != null) {
    dep = null; fn = null;
    if (d.result == null) {
        try {
            d.completeValue(f.get());   //Supplier 的 get() 方法
        } catch (Throwable ex) {
            d.completeThrowable(ex);
        }
```

```
        }
        d.postComplete();
    }
}
..
}
```

ForkJoinPool 执行上面的 AsyncSupply 对象的 run()方法，实质就是执行 Supplier 的 get()方法。执行结果被塞入了 CompletableFuture d 当中，也就是赋值给了 CompletableFuture 内部的 Object result 变量。

调用 d.postComplete()，也正是在这个函数里面，把第 2 步压入的 UniApply 对象弹出来执行，代码如下所示。

```
final void postComplete() {
    CompletableFuture<?> f = this; Completion h;
    while ((h = f.stack) != null ||
        (f != this && (h = (f = this).stack) != null)) {
        CompletableFuture<?> d; Completion t;
        if (f.casStack(h, t = h.next)) {   //出栈
            if (t != null) {
                if (f != this) {
                    pushStack(h);
                    continue;
                }
                h.next = null;
            }
            f = (d = h.tryFire(NESTED)) == null ? this : d;   //执行出栈任务
        }
    }
}
```

第 3 步：CompletableFuture future3 = future2.thenRun(…)

第 3 步和第 2 步的过程类似，构建了一个 UniRun 对象，这个对象被压入第 2 步的 CompletableFuture 所在的栈中。第 2 步的任务，当执行完成时，从自己的栈中弹出 UniRun 对象并执行。

总结一下上述过程，如图 8-5 所示。

通过 supplyAsync/thenApply/thenRun，分别提交了 3 个任务，每 1 个任务都有 1 个返回值对象，也就是 1 个 CompletableFuture。这 3 个任务通过 2 个 CompletableFuture 完成串联。后 1 个任务，被放入了前 1 个任务的 CompletableFuture 里面，前 1 个任务在执行完成时，会从自己的

栈中,弹出下 1 个任务执行。如此向后传递,完成任务的链式执行。

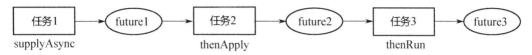

图 8-5　任务的链式串联过程

8.4.4　thenApply 与 thenApplyAsync 的区别

在上面的代码中,我们分析了 thenApply,还有一个与之对应的函数是 thenApplyAsync。这两个函数调用的是同一个函数,只不过传入的参数不同。

```
    public <U> CompletableFuture<U> thenApply(
    Function<? super T,? extends U> fn) {
    return uniApplyStage(null, fn);
}
    public <U> CompletableFuture<U> thenApplyAsync(
    Function<? super T,? extends U> fn) {
    return uniApplyStage(asyncPool, fn);
}
    private <V> CompletableFuture<V> uniApplyStage(
    Executor e, Function<? super T,? extends V> f) {
    if (f == null) throw new NullPointerException();
    CompletableFuture<V> d = new CompletableFuture<V>();
    if (e != null || !d.uniApply(this, f, null)) {
        UniApply<T,V> c = new UniApply<T,V>(e, d, this, f);   //thenApplyAsync
                                                              //的执行逻辑
        push(c);
        c.tryFire(SYNC);
    }
    return d;
}
```

最关键的是上面几行加粗的代码。

如果是 thenApplyAsync,则 e != null,构建 UniApply 对象,入栈;

如果是 thenApply,则会调用 d.uniApply(this,f,null),该函数代码如下:

```
final <S> boolean uniApply(CompletableFuture<S> a,
                    Function<? super S,? extends T> f,
                    UniApply<S,T> c) {
    Object r; Throwable x;
    if (a == null || (r = a.result) == null || f == null)
```

```
            return false;    //如果依赖的前置任务没有完成,则会直接返回 false
    tryComplete: if (result == null) {
        if (r instanceof AltResult) {
            if ((x = ((AltResult)r).ex) != null) {
                completeThrowable(x, r);
                break tryComplete;
            }
            r = null;
        }
        try {
            if (c != null && !c.claim())
                return false;
            @SuppressWarnings("unchecked") S s = (S) r;
            completeValue(f.apply(s));    //否则,执行该任务
        } catch (Throwable ex) {
            completeThrowable(ex);
        }
    }
    return true;
}
```

通过上面的代码可以看到:

(1)如果前置任务没有完成,即 a.result = null,则上面的 uniApply 会返回 false,此时 thenApply 也会走到 thenApplyAsync 的逻辑里面,生成 UniApply 对象入栈;

(2)只有在前置任务已经完成的情况下, thenApply 才会立即执行,不会入栈,再出栈,此时 thenApply 和 thenApplyAsync 才有区别。

同理, thenRun 与 thenRunAsync、thenAccept 与 thenAcceptAsync 的区别与此类似。

8.5 任务的网状执行:有向无环图

如果任务只是链式执行,便不需要在每个 CompletableFuture 里面设 1 个栈了,用 1 个指针使所有任务组成链表即可。

但实际上,任务不只是链式执行,而是网状执行,组成 1 张图。如图 8-6 所示,所有任务组成一个有向无环图:

任务 1 执行完成之后,任务 2、任务 3 可以并行,在代码层面可以写为:future1.thenApply(任务 2)、future1.thenApply(任务 3);

任务 4 在任务 2 执行完成时可开始执行；

任务 5 要等待任务 2、任务 3 都执行完成，才能开始，这里是 And 关系；

任务 6 在任务 3 执行完成时可以开始执行；

对于任务 7，只要任务 4、任务 5、任务 6 中任意一个任务结束，就可以开始执行。

总而言之，任务之间是多对多的关系：1 个任务有 n 个依赖它的后继任务；1 个任务也有 n 个它依赖的前驱任务。

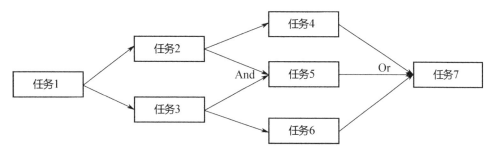

图 8-6　所有任务组成的有向无环图

这样一个有向无环图，用什么样的数据结构表达呢？And 和 Or 的关系又如何表达呢？

有几个关键点：

（1）在每个任务的返回值里面，存储了依赖它的接下来要执行的任务。所以在图 8-6 中，任务 1 的 CompletableFuture 的栈中存储了任务 2、任务 3；任务 2 的 CompletableFuutre 中存储了任务 4、任务 5；任务 3 的 CompletableFuture 中存储了任务 5、任务 6。也就是说，每个任务的 CompletableFuture 对象的栈里面，其实存储了该节点的出边对应的任务集合。

（2）任务 2、任务 3 的 CompletableFuture 里面，都存储了任务 5，那么任务 5 是不是会被触发两次，执行两次呢？

任务 5 的确会被触发 2 次，但它会判断任务 2、任务 3 的结果是不是都完成，如果只完成其中一个，它就不会执行。

（3）任务 7 存在于任务 4、任务 5、任务 6 的 CompletableFuture 的栈里面，因此会被触发三次。但它只会执行一次，只要其中 1 个任务执行完成，就可以执行任务 7 了。

（4）正因为有 And 和 Or 两种不同的关系，因此对应 BiApply 和 OrApply 两个对象，这两个对象的构造函数几乎一样，只是在内部执行的时候，一个是 And 的逻辑，一个是 Or 的逻辑。

```
    BiApply(Executor executor, CompletableFuture<V> dep,
        CompletableFuture<T> src, CompletableFuture<U> snd,
        BiFunction<? super T,? super U,? extends V> fn) {
    super(executor, dep, src, snd); this.fn = fn;
}
    OrApply(Executor executor, CompletableFuture<V> dep,
        CompletableFuture<T> src,
        CompletableFuture<U> snd,
        Function<? super T,? extends V> fn) {
    super(executor, dep, src, snd); this.fn = fn;
}
```

（5）BiApply 和 OrApply 都是二元操作符，也就是说，只能传入两个被依赖的任务。但上面的任务 7 同时依赖于任务 4、任务 5、任务 6，这怎么处理呢？

任何一个多元操作，都能被转换为多个二元操作的叠加。如图 8-7 所示，假如任务 1And 任务 2And 任务 3=任务 4，那么它可以被转换为右边的形式。新建了一个 And 任务，这个 And 任务和任务 3 再作为参数，构造任务 4。Or 的关系，与此类似。

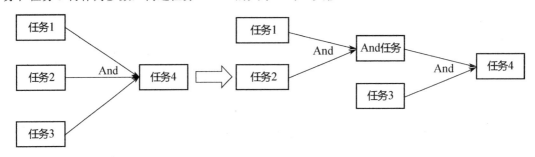

图 8-7 一个多元操作转换为多个二元操作

明白了任务的有向无环图的存储与计算过程，也就明白了 8.1.4 节 thenCombine 的内部实现原理。thenCombine 用于任务 1、任务 2 执行完成，再执行任务 3，实际场景更为简单，此处不再进一步展开源码讨论。

8.6 allOf 内部的计算图分析

下面以 allOf 函数为例，看一下有向无环计算图的内部运作过程：

```
public static CompletableFuture<Void> allOf(CompletableFuture<?>… cfs) {
    return andTree(cfs, 0, cfs.length - 1);
}
static CompletableFuture<Void> andTree(CompletableFuture<?>[] cfs,
```

第 8 章 CompletableFuture

```
                                    int lo, int hi) {
    CompletableFuture<Void> d = new CompletableFuture<Void>();
    if (lo > hi) // empty
        d.result = NIL;
    else {
        CompletableFuture<?> a, b;
        int mid = (lo + hi) >>> 1;
        if ((a = (lo == mid ? cfs[lo] :
                  andTree(cfs, lo, mid))) == null ||
            (b = (lo == hi ? a : (hi == mid+1) ? cfs[hi] :
                  andTree(cfs, mid+1, hi)))  == null)
            throw new NullPointerException();
        if (!d.biRelay(a, b)) {
            BiRelay<?,?> c = new BiRelay<>(d, a, b);
            a.bipush(b, c);    //会把c分别压入a, b所在的栈中
            c.tryFire(SYNC);
        }
    }
    return d;
}
```

上面的函数是一个递归函数，输入是一个 CompletableFuture 对象的列表，输出是一个具有 And 关系的复合 CompletableFuture 对象。最关键的代码如上面的加粗代码所示，因为 c 要等 a, b 都执行完成之后才能执行，因此 c 会被分别压入 a, b 所在的栈中。

```
final void bipush(CompletableFuture<?> b, BiCompletion<?,?,?> c) {
    if (c != null) {
        Object r;
        while ((r = result) == null && !tryPushStack(c))   //c压入a的栈中
            lazySetNext(c, null); // clear on failure
        if (b != null && b != this && b.result == null) {
            Completion q = (r != null) ? c : new CoCompletion(c);
            while (b.result == null && !b.tryPushStack(q))    //c压入b的栈中
                lazySetNext(q, null); // clear on failure
        }
    }
}
```

图 8-8 所示为 allOf 内部的运作过程：方块表示任务，椭圆表示任务的执行结果。假设 allof 的参数传入了 future1、future2、future3、future4，则对应四个原始任务。

生成 BiRelay1、BiRelay2 任务，分别压入 future1/future2、future3/future4 的栈中。无论 future1 或 future2 完成，都会触发 BiRelay1；无论 future3 或 future4 完成，都会触发 BiRelay2；

生成 BiRelay3 任务，压入 future5/future6 的栈中，无论 future5 或 future6 完成，都会触发 BiRelay3 任务。

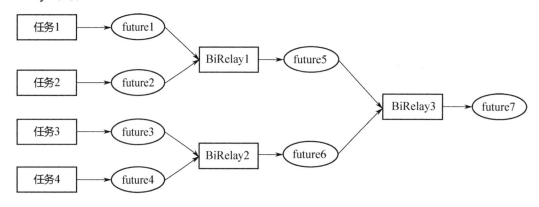

图 8-8　allOf 内部的运作过程

BiRelay 只是一个中转任务，它本身没有任务代码，只是参照输入的两个 future 是否完成。如果完成，就从自己的栈中弹出依赖它的 BiRelay 任务，然后执行。